Advances in Spatial and Network Economics

Titles in the Series

Martin J. Beckmann and *Tönu Puu*
Spatial Structures

Aisling J. Reynolds - Feighan

The Effects of Deregulation on U. S. Air Networks

With 15 Figures

Springer-Verlag

Berlin Heidelberg New York
London Paris Tokyo
Hong Kong Barcelona
Budapest

Dr. Aisling J. Reynolds-Feighan
Department of Economics
Transport Policy Research Institute
University College Dublin
Belfield, Dublin 4
Ireland

ISBN-13: 978-3-642-77063-0 e-ISBN: 978-3-642-77061-6
DOI-13: 10.1007/ 978-3-642-77061-6

2142/7130-543210 - Printed on acid-free paper

To Kieran

ACKNOWLEDGEMENTS

I would like to thank the following people for their helpful advice and comments: my doctoral advisor, Geoffrey J.D. Hewings, and James Huff, Arthur Getis and Charles Kolstad at the University of Illinois; David Batten and Åke Andersson who provided many useful suggestions; Douglas K. Flemming for putting me in touch with the staff at Continental Airlines; William Brunger, Tim Griffin and particularly Larry Chestler at Continental Airlines who provided me with the extensive data set used in this book; my husband, Kieran Feighan for his perceptive comments and advice all the way along.

I would also like to thank my family for their support: Dan and Rita, Enda, Justin and Gwenville.

Financial support was provided in part by the Regional Economics Applications Laboratory (REAL) at the University of Illinois and the Federal Reserve Bank of Chicago, and by the Ameritech Corporation.

CONTENTS

LIST OF FIGURES

LIST OF TABLES

Table	*Title*	*Page*

CHAPTER 1
INTRODUCTION

1.1 A Brief History of U.S. Commercial Aviation Regulation and Deregulation

The U.S. commercial aviation industry was regulated by the government for a period of 40 years, beginning in 1938 with the passing of the Federal Aviation Act, and ending in October 1978 when President Carter signed into law the Airline Deregulation Act (ADA). There were 16 airlines in existence when the Federal Aviation Act was passed in 1938 (the so-called 'trunk lines'). The Act established the Civil Aeronautics Authority (CAA) as the industry's regulatory body. The Act was passed principally because it was felt that the free market, if allowed to continue unregulated for much longer, would put many of these firms into bankruptcy. It is possible therefore to view the CAA of 1938 (re-organized into the Civil Aeronautics Board (CAB) in 1940) as a response to a potential market failure at the time. In the 1930s, few air traffic markets could have efficiently supported more than one airline operating in the market [Panzar (1980)]. Competition among the carriers was cut-throat, and it was felt that the near bankruptcy of the airlines in the period was caused principally by the competitive bidding system used by the Post Office in allotting airmail subsidies [Keeler (1972), Caves (1962)]. Given that for most routes it was economical for only one carrier to transport the mail, and impossible for an airline to be viable carrying passenger traffic alone, airline markets can be considered to have been joint product natural spatial monopolies. It was therefore considered sound economic policy, to consolidate the responsibility for determining mail rates and passenger fares into a single entity [Panzar (1980)]. The CAA/CAB controlled the air carriers in the industry, while the Federal Aviation Administration (FAA) operated and controlled the airport system as well as promulgating and enforcing safety regulations.

The CAA/CAB was given authority over routes, entry into the industry and entry into markets. In addition to controlling entry by new firms, they also restricted the number of existing carriers that would compete in a given market. The trunk airlines were permitted to serve the major long-haul markets, while other airlines, which entered the industry after the passage of the Act, were permitted only to serve regional markets. Accordingly, a degree of route and market specialization began to develop [Graham, Kaplan and Sibley (1983)].

The trunk carriers were commonly assumed to serve the more profitable routes; they were therefore ineligible for subsidies. Local carriers (post-1940s and 1950s entrants) were presumed to serve less profitable routes than the trunk carriers. Until the 1960s, the CAB severely restricted these carriers' ability to compete with the trunks so that, for example, the route certificates included explicit prohibitions against providing nonstop service in markets already served by the trunk airlines. Regulation emphasized the development and maintenance of the so-called 'scheduled services'[1]. The scheduled service carriers assumed the obligation of service and had made long-term investments to provide constant service; while they commenced routes voluntarily, the obligation to serve at all times was negotiated with government regulators. In exchange for a service obligation, such carriers received protection in the form of restriction of entry by other carriers, an opportunity to charge minimum rates and/or sanctioned price discrimination.

Without entry control in the industry, the possibility of non-scheduled carriers entering only the more lucrative markets was high. This would leave the thinner markets to the established scheduled carriers. Since charter fares would have a downward impact on scheduled fares, the reduced returns experienced by the scheduled carriers would erode the internal subsidies that the high volume routes provided for the thin markets. Accordingly, without entry and price restriction, the thinner markets would lose service. The claim was, therefore, that regulation allowed

[1]Scheduled services are defined [Kamp (1976)] as those services that operated on the basis of a published timetable open to the general public, available at a published and fixed price and governed by a contractual relationship between the passenger and the air carrier.

for service to a much larger number of cities than unregulated services would allow, and accordingly reduced the need to subsidize small-town service with tax revenues. Questions were asked about the amount of cross-subsidization that was actually taking place in the 1960s and 1970s, as well as the fact that commuter/air taxi carriers (which were unregulated by the CAB) had managed to provide profitable (unsubsidized) service on many of the routes on which even the subsidized local service carriers had not been able to break even [Keeler (1972)].

Through the 1960s and 1970s, criticisms of the CAB's regulatory regime grew. In the U.S., regulators had refrained from implementing radical policy changes since the early days of creating a scheduled air transport system. By the 1960s, changes in demand, increases in fuel prices and changes in the economy meant that a more entrepreneurial approach to regulation was required in the western world. With entry restrictions and price competition eliminated via regulation, the focus shifted to equipment and scheduling competition. It is suggested that the change created a continuous overcapacity problem for some airlines. This problem was made worse by slackening demand in times of economic stagnation and misjudgments of costs, issues that became increasingly visible in the 1970s due to inflationary pressures and increased fuel expenses [Moore (1986), Kamp 1976)]. The regulated scheduled air transport industry, it was claimed, fulfilled a role in the nation's defense system and enabled economic development in remote areas, thereby serving the same function as most other elements of transportation or other parts of a nation's infrastructure. Safety was a further reason given for airline regulation, and questions were posed as to why fares and entry should have been controlled, given the FAA's strict safety standards [Keeler (1972)].

The CAB came to believe that the marketplace would be more effective than the Board in determining the markets that carriers served and the prices they charged. In 1977, the CAB made moves to reduce entry barriers to the airline industry and also to permit pricing freedom. In 1978, Congress acted to eliminate many controls over the airlines. The Airline Deregulation Act (ADA) of October 1978 codified the earlier CAB actions and set out a six year transition period for the Board to implement all the requirements of the new law. A series of programs were instigated by the CAB in order to meet the ADA requirements. These requirements were aimed at

(1) increasing the air transportation industry's reliance on competition,

(2) increasing the industry's reliance on private sector forces,

(3) reducing Federal involvement and intervention, and

(4) stimulating investors, new competitors and community action [CAB (1984)].

The Act instructed the CAB to grant operating rights to any air carrier applying to serve a route on which a specified minimum level of service was not being provided, even though other carriers had authorizations to provide the service. 'Essential air transportation services' were to be provided for a ten year period and provision was made for subsidies to cover any losses. Without CAB approval, carriers were free to vary prices by as much as 50% below or 5% above the standard industry fare. The CAB's authority over routes was to end in December 1981. By the end of 1982, its authority over rates, mergers and acquisitions was to cease, with the CAB's own demise scheduled for December 1984 [Moore (1986), CAB (1982, 1984), Meyer, Oster, Morgan, Bergman and Strassman (1981)].

The Board rapidly implemented the terms of the Act. New routes were freely awarded and the gradual withdrawal from any regulatory action was observed. Before the 1981 deadline, entry was reasonably free for any firm able to establish its fitness. The Board had allowed airlines to price with almost total freedom long before its loss of authority to control rates.

Generally speaking, the industry has become more competitive since 1978. There is an increase in the number of carriers serving particular market segments and market concentration within the industry has been reduced [CAB (1984)]. The legal and policy standards, under which mergers and acquisitions were decided, have also changed substantially since 1978. Congress directed the CAB to deal with airline mergers in a manner similar to that governing standards for unregulated industries and, accordingly, placed airline mergers in the same category as those in other industries.

As a result of this change, the Board's primary concern (now the FAA's) was to prevent combinations which would substantially lessen competition or tend to create a monopoly at particular airports, or more importantly, on particular routes [CAB (1982)].

The Board instigated a series of programs aimed at making a smooth transition to the deregulated environment. The CAB in this way met the ADA's requirements to increase the air transportation industry's reliance on competition and on private sector forces, to reduce Federal involvement and intervention and to stimulate investors, new competitors and community action [CAB 1984)].

1.2 Critique of Other Studies

Given that just over a decade has passed since the passage of the Act, and that by now the market is considered free from regulation (at least free in the sense that Congress had envisaged it to be), a general assessment of the impact of this legal reform can be made, since many of the transitional patterns and upsets have died out at this stage. The deregulation of the U.S. airline industry had a major impact on the organization, size and operation of that industry. One of the most significant changes brought about by passage of the Act was the reorganization, growth and expansion of the airlines' networks. This key element in the airline's operation has received relatively little attention in assessments of deregulation in the last decade. The impact of the ADA on airline networks forms the principal focus for this book. In the next chapter, a review is given of the impact of the ADA as made by economists and planners studying the industry. This chapter illustrates the focus by economists on industry structure and organization, production factor performance and productivity and consumer effects and welfare changes induced by passage of the Act. The views of transport planners, engineers, geographers and regional scientists are then put forward and the key role of the airline's network in the production of flight services is explained.

The CAB suggested that the carrier's strategy for developing a 'competitive niche in the deregulated industry' depended very much on its status before 1978 [CAB (1984)]. Certainly the number of airlines greatly increased in the period immediately following deregulation. Newcomers to the industry then tended to be low-cost, non-union airlines which initially had a significant cost advantage over the trunks. These new airlines typically served short-haul markets, or markets where the proportion of discretionary passengers was high. The main focus for these new carriers was on low fares, but as the trunks gradually began to match their prices, competition based on quality of service (load factor, inflight services and the ratios of flight attendants to passengers, delay factors etc.) became evident again. The total number of carriers serving the industry in 1988/89 has fallen significantly from the numbers in 1980/81. The market under free entry and exit now resembles the market that existed under regulation, in terms of numbers of carriers and market shares. In Chapter 3, the U.S. Airline industry is profiled over a 20 year period from 1969-88. In this chapter, the emphasis is very much on the spatial characteristics of the National Airways System (i.e. the system of commercial airports in the U.S.) and the spatial organization of production of flight services by airlines before and after 1978. The impact of air carrier network changes on the airport system is assessed and quantified, with the traffic patterns and network organization of two major U.S. carriers being examined in detail. Statistical measures which suitably capture the pertinent characteristics of an airline's network are evaluated and used to assess the extent of network changes by the two U.S. carriers.

After 1978, the existing carriers expanded their route systems with great emphasis on the so-called 'hub-and-spoke' network system. While the large trunks have experienced some decline in their market shares, they have tried to maintain their positions in the deregulated environment by rationalizing their route structure and making more widespread online connections available to their passengers. The smaller trunks have contracted, while local service carriers have rapidly expanded. Indeed it was the local service carriers that first established hub-and-spoke operations in the early 1970s, since a large proportion of their passengers connected to trunk carriers. Their success stimulated the trunks to experiment and develop hub-and-spoke systems, [CAB (1984), Koran (1983), Moore (1986), Meyer (1986)] and Chapter 3 contains further details of these trends and industry changes.

1.3 A Production Model of the Airline Firm

In microeconomic theory until recently, analysis of the firm has focused particularly on the single-product firm. Recent literature has placed emphasis on the development of theory and empirical models of multiproduct firms, where one firm produces a variety of products. An airline company is one such firm, producing flights over a large number of routes in its network. The complexities involved in dealing with the analysis of an airline firm compared to a simple one-product firm will be focused on throughout this book. Looking at the one-product firm firstly, its goal is assumed to be the maximization of its profit. It therefore needs to determine what prices to charge for its outputs or pay for its inputs, and what levels (or quantities) of outputs or inputs to produce or use. The firm will make these decisions against two broad kinds of constraints. Technological constraints determine the feasibility of a production plan, while market constraints concern the effects of other agents, for instance consumers and suppliers of inputs. For a competitive firm, the most simplistic type of economic behavior is assumed, namely that all prices are given, so that the firm can only make decisions about the quantities of outputs and inputs to carry.

The firm will typically face a constrained profit maximization problem in a given time period, and under these circumstances there are two possible single-decision choices for the firm: on one hand, the firm can maximize the output for a given cost or budget (therefore, the cost function is fixed), or, on the other, it can minimize costs of producing a given level of output (in which case the production function is fixed). The production function is a technical relationship, relating factor inputs to output: it describes the technically efficient methods for the transformation of these inputs into output (or products) at a particular point in time. This function may be formulated to represent the technology of a firm, an industry or indeed an economy. The cost function gives the minimum cost of producing a level of output X, when factor prices are given.

The estimation of production functions is useful because it explicitly sets out the technical relationship between inputs and outputs for a firm or an industry. Having estimated the function, the marginal productivity of the factors, the marginal rates and elasticity of substitution may be calculated, along with an assessment of the efficiency of production and the nature of returns to scale for the firm or industry. The production problem and cost problem are intimately related and by estimating the production function, statements about the firm's cost relationships may be made.

When examining firm or industry performance in most industries, it is obvious that firms typically produce more than one product, so that analysis using traditional microeconomic analysis fails to realistically capture 'real world' production processes. Traditional analysis simplifies what is a complex process for most firms, as well as simplifying the market environment in which these firms operate.

In the 1980s, the work of Baumol, Panzar and Willig (1982) did much to focus firm and industry analysis on the multi-product firm. The literature since then has begun to address the basic characteristics and production relationships of these firms as well as looking at concepts of efficiency and performance under these new circumstances[2].

The multi-product firm produces a series of differentiated products and in many instances there are 'economies of scope[3]' associated with producing products which either (a) share a fixed input factor or (b) have one input that has properties of a public good so that it is available for other production processes even though it has been purchased for production of one specific product, or (c) are connected with a network of operations of one

[2]See Bailey and Friedlaender (1982) and Baumol and Willig (1986) for reviews on progress in this area.

[3]Economies of scope exist when a single firm can produce a given level of output of each type of product more cheaply than a series of separate firms each producing a given level of output of one product [Bailey and Friedlaender (1982)].

sort or another (e.g. communications, information or transportation networks). Airlines fall into this latter category, where the production of flight services on one route influences production of services on other routes by generating increased passenger flows between routes. Because there is a larger scope of operations, "economies of scope of a specialized input, airplanes, can be reaped" [Bailey and Friedlaender (1982)].

Each route in the airline's network represents a different product and the characteristics of these routes need to be included in the production function formulation in a clear, coherent manner. While some attempts have been made by economists to include measures which try to capture the nature of the network at the firm or route level [Borenstein (1989); Kanafani and Hansen (1985); Morrison and Winston (1990)], these have tended to be randomly selected variables which have no sound theoretical basis for inclusion. In Chapter 4 of this book, a model of the individual airline firm is put forward explicitly including such a network variable in the firm's production function. In this Chapter, the model is developed and the precise form of the network variable is explained. The joint-product nature of the production process implies that there exists an association or dependence between potentially all routes in the network system. In developing the empirical production model for the airline firm, this dependency is captured by estimating a spatial autoregressive model, where the connection rates of passengers between the different routes are used as a measure of the dependency in the system. In order to test the model, detailed route by route data were obtained from a major U.S. carrier for the period 1987/88. In Chapter 5, these data are described in detail as are several additional variables that were generated for the airline firm model.

1.4 Measuring Technical Efficiency in an Airline Network

The production function explained above, represents a frontier in the sense of setting a limit to the range of possible output levels. Production levels falling short of this limit can be considered relatively inefficient, so that the degree to which a firm's output levels lie below it's production function may be regarded as a measure of inefficiency. Inefficiency may occur for technical reasons (i.e. employing excessive inputs to produce a given level of output), for allocative reasons (employing inputs in the wrong proportions) or for scale reasons (where the firm is operating at a scale where price is not equal to marginal costs). Since the analysis in this book focuses on production relations, the main concern is with technical inefficiency: for the airlines, a problem that occurs when too many inputs are being used to produce a given level of output on a particular route, given the interconnections between routes.

Econometricians estimate the various frontier or 'best practice' functions by recasting the theoretical framework in such a manner as to allow for calculation of the magnitudes and costs of the various types of inefficiency relative to these frontiers. In Chapter 6, the new production model is estimated and compared with 'best practice' functions so that the relatively more efficient routes may be identified and the characteristics of these routes highlighted. In this way the technical efficiency of a hub-and-spoke network system may be evaluated and general conclusions about this form of network organization may be made.

In the final chapter, the conclusions to the study are set out, along with recommendations for further investigations. The over-riding theme is that airline networks represent both the product of the firm and also the production plan: each segment may be seen as an asset of the firm, contributing to the production of flight services not only on that route, but also on adjoining routes, since connecting passengers are 'byproducts' in the process. The structure of networks has changed substantially since 1978, although it represents the logical progression from their regulated form. As such, the traffic distribution at U.S. airports has changed dramatically since deregulation. Emphasis throughout the book is on this key component in the firm's operation, and how it may be viewed at the industry and national level.

In reviewing the academic literature on airline deregulation, it is clear that economists and geographers on one hand tend to analyze and study the industry and the airport network at a macrolevel. On the other hand, the scale of analysis for operations researchers and transport planners and engineers is necessarily at a microlevel, since researchers in these fields concern themselves with solving specific problems related to the individual airline's short-term operations. The formulation of such problems gives a terrific insight into what those working in the industry see as being the most important components of the airline's success and profitability. Many of the problems and shortcomings in the economic analysis of the airline industry can be resolved to some extent by viewing the problem at a different scale and realizing the role of an important characteristic like network organization at the micro-level. This information may then be used to deal with analysis at a more aggregated level. Very substantial evidence is presented in Chapter 3 as to the appropriateness of reviewing how airline operations need to be modeled.

In formulating the model of the airline, the firm's decision-making process is viewed from a production theory framework rather than from a cost theory framework. There are two principal reasons for this. First, it is not possible to obtain detailed information on airline input and output prices/costs. Given the highly competitive nature of this industry, no firm is willing to allow detailed analysis of its economic structure, on a market by market basis. Even in formulating a production model using input and output quantities, the airlines are either not willing or not able to supply information on the use of basic factors of production. In estimating input quantities used on routes, use has to be made of company aggregated or averaged statistics. While this approach presents many problems, it would be impossible to attempt to estimate input costs/prices given the greater range of distribution of costs among categories of labor for example. The second reason for focusing on a production formulation relates to the nature of the network. The notion of a network is more easily conceptualized and specified as a physical or technological factor. As such it was decided to focus purely on the modeling of the firm's production relationships over a network.

In the next chapter, the contributions of researchers in

economics, regional science, geography, transport planning and operations research are reviewed to highlight the differing points of view and areas of concern for those analyzing the U.S. airline industry. Building on this review, the case is made for a new economic model of the airline firm.

CHAPTER 2
PREVIOUS STUDIES ON U.S. AIRLINE DEREGULATION

This chapter is set out in three sections and reviews different aspects of the literature on U.S. airline deregulation. In the first section, the issues and concerns raised by economists are summarized and some of the major findings are reported. In the second section, the small body of work done by geographers and regional scientists in the area of commercial aviation is outlined. Section 2.3., the final section, deals with research undertaken by operations researchers in transport and can generally be described as network or scheduling studies. The emphasis in this last section is placed on the conceptualization and formulation of the problems rather than on solution methods and detailed empirical findings.

2.1 Economic Research on U.S. Airline Deregulation

Academic research in economics focusing on the airline industry has consistently viewed problems and made assessments at the industry level; the individual firm has not been rigorously assessed. Research carried out prior to 1978 helped to make the case for deregulation, and economists since 1978 have been eager to examine predictions made, and quantify the gains to the firms and to the consumers that the ADA made possible. Summary data on each airline's inputs and input prices are typically used to make the assessments; one of the problems however has been the failure to see the firm's network as an important input, and indeed as an asset in production of flight services. The spatial restructuring of the U.S. airways system instigated by carriers' propensity to significantly streamline their passenger flows through their networks has not received any attention in the literature. The specific nature of airlines' network changes has not been examined satisfactorily, and accordingly assessments made of the post-ADA patterns are lacking.

In the economics literature prior to deregulation, academic research tended to take one of two points of view. On one hand, some writers questioned the need for any governmental economic regulation of the airlines, [Eads, Nerlove and Raduchel (1969), Devany (1975), Verleger (1972)] while on the other, attempts were made to build analytical models of industry behavior under price and entry regulation, and so derive more enlightened regulatory policy recommendations [Douglas and Miller (1974)]. Theoretical analysis of unregulated airline markets received little attention at that time [Panzar (1980)].

In the first approach, analysis of costs and demand for air services formed the major part of the research. Demand for air transportation under price and entry regulation was typically examined with firms varying the quality of their product [Devany (1975), Verleger (1972), White (1972)].

By incorporating a measure of service in a model of airline demand, it was demonstrated that quality of service variables played a strong role in the regulated period, with airline demand being positively related to the number of flights and negatively related to the load factor [Ippolito (1981)]. Comparison of quality in regulated competitive and monopolistic markets formed the basis for another paper [White (1972)]. White demonstrated, using several airline demand models, that the competitive firm offers a higher level of quality under regulation than does the monopolist in the same circumstances. He shows that in general, regulation produces uniformity in the level of quality offered compared with a non-regulated industry, where a variety of quality levels are offered to consumers, increasing their choice set. Another study concentrated on the appropriate method of estimating demand functions in regulated markets, and one conclusion reached here was that cross-sectioned models are inadvisable for analyzing the effects of regulatory policy. This result obtains because the entire U.S. air traffic market is made up of a diverse set of heterogeneous markets, and accordingly analysis is "impossible" using cross-sectioned techniques (such as unconstrained gravity model formulations) to analyze this system [Verleger (1972)]. Verleger proposes that account be taken of regional differences in income and population for example, when undertaking point-to-point analysis, but he does not address the form that these models should take. Fotheringham's work (1983) suggests that in fact separate parameters be estimated for each node in the system to take account of these variations. In time series models however, the nature of the network is usually ignored so that the technology of network organization is assumed to be either constant over time or of little importance. This view is disputed in this book and the important role of the network in both time series and cross-sectioned studies is clearly demonstrated.

The second approach mentioned above aimed at finding optimal fare and quality of service levels under a regulatory regime by modeling the regulated market. The argument was put forward that price controls in a monopoly situation affect technical efficiency and production of a specifically defined output, while in non-monopoly markets, such controls affect the quality level of the output and is therefore an issue of allocative efficiency. Accordingly, by selecting price levels, the regulators indirectly control quality since the regulated price is related to the relative excess supply associated with the equilibrium. In order to examine airline costs and make an efficient price level selection, it was argued that the relationships among slack capacity, passenger costs and overall service quality needed to be assessed. By assessing these relationships, optimal air fare levels and structures were presented, along with a comparison of these optimal levels vis-à-vis prevailing levels under regulation. The general conclusion was that costs among regulated carriers were high because the imposed price level was high and not the reverse [Douglas and Miller (1974)].

In the early 1980s, the notion of a 'contestable' market was put forward in the economics literature [Baumol (1982)]. Essentially, a perfectly contestable market is one in which a necessary condition for an industry configuration to be in equilibrium is that it be sustainable (i.e. profits are at least zero for all firms supplying the market and negative for potential entrants). At equilibrium, the cost minimizing number of firms (a 'feasible industry configuration') produce output at a certain price level. This framework makes market structure an endogenous variable, and is a more general theory than perfect competition. The hypothesis puts regulators in a poor light, since it questions the desirability of artificial impediments to entry and the propensity of regulators to resist the closing down of unprofitable lines of activity [Baumol (1982)]. In many ways, this hypothesis represents a return to more traditional economic modeling of industrial organization. The hypothesis suggests a static model of competition, and accordingly does away with much of the richness that models utilizing a 'game-theoretic' approach have to offer. In the late 1970s and early 1980s, the airline industry, which was thought to represent a good example of a contestable market, was assessed in this light along with a more general attempt to quantify the degree of change for producers and consumers in the regulated environment.

The contestability hypothesis is not supported by researchers in the literature. It was found for instance, that increasing the number of firms beyond two does seem to have an effect upon price, a result inconsistent with the theory, since the argument posits the view that potential competition would keep fares at cost even in highly concentrated markets [Graham et al. (1983)]. Moore (1986) suggests that the fare reductions and expected effects of contestability occur when the number of firms in a given market is five or more. In contrast, Trapani and Olson (1981) examine the effects of market concentration on capacity and the quality of service. They show that the competitive airline market at the optimal solution will have a lower fare and lower quality of service than the previously regulated counterpart and, that the estimated change in market concentration required to bring about the optimal solution is relatively small.

Reiss and Spiller (1989) have analyzed the nature of competition and entry in small airline markets, concluding that firms' competitive conjectures differ across competing segments (i.e. routes), and, that demand and cost parameters differ depending on the extent of cross-service competition. The authors argue against aggregating direct and indirect services when studying market power. In their analysis of profitability, the authors implicitly assume that a carrier's variable profits on one route are not affected by the carrier's operations on other routes or networks. In the analysis of multiproduct firms however, one of the key arguments put forward is that the existence of economies of scope in an airline network forcefully encourages firms to produce over a network, rather than point-to-point operations. So surely carrier operations on one route do influence operations on other routes, to the extent that a seemingly inefficient or unprofitable route may, in fact be considerably more efficient/profitable when viewed as part of an integrated system. Again the failure to treat the network in an explicit and formal way has lead to many contradictions or disputes in the literature assessing various aspects of the airline industry's performance before and after deregulation.

In terms of input factor performance and productivity, it is suggested that capital has benefited from deregulation, while organized labor has most likely lost since the stock value of airlines over the last 8 years has remained steady, with most gains accruing to the regional carriers. In the labor market, while total employment was up in the 1976-82 period, some erosion of wage rates was evident with the cost-minimizing consciousness of the new airlines weakening the stance of organized labor in the older, established airlines [Moore (1986), Morrison & Winston (1986)].

In terms of total factor productivity growth of the airlines for the period 1976-80, the U.S. airline industry attained very rapid growth in output and productivity. In 1980 however, output and productivity decreased at very high rates. The growth in productivity is attributed to two main sources. The first and most important comes from changes in readily quantifiable operating characteristics (such as level of output, load factor, average stage length and available capacity), most especially output. Secondly, residual or unexplained productivity growth can occur due to changes in regulation, managerial efficiency and other factors that are "difficult to quantify" [Caves et al. (1983)]. The effects of network organization would, no doubt, be included in this latter category and it will be shown that this factor in particular represents not only a key component of technological change in airline operations but also a dramatic change in the airlines' products.

Bailey, Graham and Kaplan, (1985) in contrast, do not find a trend toward accelerating productivity growth for the trunk lines in the 1976-81 period. The suggestion is made that any deregulation-induced productivity gains during the 1978-81 period were apparently dominated by other factors. Again the network potentials for each route are not considered as inputs in the production of flight services, and also the joint-product nature of the airlines' products is not satisfactorily understood. Given this omission at the firm level, it is not surprising that, at the industry level, there is even less understanding of, or catering for, the network organization.

Caves et al. (1984) formulate a general model of airline costs and proceed to examine economies of density (as opposed to scale or scope) in an attempt to explain why it is that trunk and local service airline costs differ. The authors emphasize the role of the airline network in their model and use the number of points served as their measure of the size of each carrier's service network. The main conclusion drawn is that density of traffic within an airline's network is the major factor explaining cost differences; the role of average stage length is of secondary importance. While counting the number of points served does provide a measure of scale for each airline, it does little to address the issue of organization of service over those points, and therefore the cost effectiveness of the airline's production plan.

Morrison and Winston (1986) attribute the $2.54 billion gain in industry profits to deregulation, and claim that the poor industry performance in the early 1980s would have been worse had regulations remained in effect. They claim that deregulation enabled carriers to increase load factors and flights through flexibility in pricing and improved scheduling practices and to improve the use of inputs through more efficient use of hub-and-spoke routing. The deregulation-induced increases in output and load factors, as well as decreases in the growth of capacity, the authors conclude, have significantly increased productivity growth in the airline industry.

Looking at deregulation from the consumer's viewpoint, it is claimed that for passengers, the quality of service has diminished since deregulation, though it is pointed out that it is the passengers who have chosen the lower-quality service with lower rates over higher quality with higher fares [Moore (1986)]. It is suggested that overall, there is a gain to the consumer resulting from the fall in fares which is sufficient to offset the deterioration of service quality, assuming some price numeraire to relate the two [Koran (1983)]. The improved product offered to the consumer in the form of superior network organization leading to better on-line connections and flight frequencies are not considered in the equation. Producer surplus, it is argued, has remained substantially unchanged, so that in general the social welfare effect has been positive.

Morrison and Winston (1986) in a comprehensive welfare assessment of the ADA quantify gains and losses to consumers, producers and society in terms of savings made by the Federal Government. They claim that the predictions made by Douglas and Miller (1974), (namely, that airline deregulation would benefit travellers because fares would fall, even though quality would decline) are largely inaccurate. The falling fares and service quality, they claim are encountered primarily in pleasure travel between large hubs. The benefits from deregulation accrue mainly to business travelers because of improved service and convenience attributable to the accelerated development of hub-and-spoke operations and to frequency improvements in low density markets. They fail however, to explicitly test the influence of the routing strategy between 1977 and 1983, their chosen testing periods. They claim that on average, all travelers have gained from fare reductions, when examined by 'distance-traveled' classifications. Given that there has been a reduction in administrative costs to the Federal Government for salaries and office space etc., and also, a decline in the cost of federally subsidized air services, society has gained in general from deregulation. The authors claim that deregulation has led to a yearly welfare gain to both travellers and carriers of roughly $8 billion (in 1977 dollars) without generating any substantial losses to specific groups in society.

It is argued that fare structure has changed dramatically, with short-haul markets increasing fares and long haul market fares decreasing, since previously, the structure of fares did not match the structure of costs. The CAB fare structure was based purely on a linear distance relationship, and while this changed after deregulation, distance-based pricing may be observed again recently. The hypothesis that CAB regulation caused airlines to employ excess capacity relative to the capacity that would have prevailed under unregulated competition is confirmed [Graham et al. (1983)]. The decrease in the growth of capacity is a factor contributing to increased productivity growth in the industry [Morrison and Winston (1986)]. Presumably better scheduling and networking by the airlines has generated greater passenger potential in all markets leading to cost savings as load factors are reduced. This view is supported by Bailey et al. (1985). They find that trunk load factors increased fairly steadily between 1974 and 1979, but fell sharply in the period 1979-81. During the latter period however, load factors were still substantially higher than they were during the previous two recessions of 1974-75 and 1969-70. The suggestion is offered that, whereas airlines were induced to offer excessive numbers of flights on long-haul routes under CAB-imposed fare structures, deregulation led to a more economically efficient distribution of service levels.

Generally speaking, airline networks have become more efficiently organized along with the changing competitive structure of the industry [Graham et al. (1983)]. However, little analysis of reorganization of the networks has appeared to date. Deregulation, it is argued, has prompted experimentation with different groups of strategies. These include hub strengthening, extension of the network via internal growth or merger/acquisition, regional rationalization and seasonality reduction [Meyer et al. (1981)].

Kanafani and Hansen (1985) argue that there is evidence of the move to hubbing by the airlines, apparent from visual comparison of current route maps with those of the mid-1970s. They also point out that measures of concentration of an airline's operation can serve as measures of the degree of hubbing for the airline's route network. Alternatively, they suggest that it should be possible to use measures for assessing industrial concentration and analyzing income distribution to measure the degree of concentration over an individual airline's network. For their study, the authors use the Herfindahl index, and an index constructed from the ratio of an airline's total departures to that airline's departures from the one, two, three or four 'most departure airports'. In addition, they claim that hubbing is evident from increased concentration of operations, more direct flights between small and large airports and greater proportion of passengers receiving on-line service. The authors argue that hubbing can increase productivity by allowing airlines to use larger, more economical aircraft, while maintaining an acceptable level of service frequency. The airlines, accordingly, can maintain higher load factors. Furthermore, hubbing allows airlines to avoid frequent costly stops necessitated by hedgehopping. The authors point out, however, that while the geographic concentration of operations at a single airport entailed by hubbing may increase productivity, the temporal concentration can have the opposite effect, so that on balance, "there is no guarantee that hubbing improves the quality of service experienced by passengers" [Kanafani and Hansen (1985), 10]. They cannot quantify or empirically support these different effects of hubbing. Furthermore, in their analysis of airline costs (which includes the Herfindahl index as a measure of hubbing) they conclude that airlines with strongly hubbed route systems incur roughly the same cost to provide a given amount of transportation as those with less hubbed systems, controlling for other factors.

In sharp contrast to this, McShan (1986) also focuses on the economic impact of the hub-and-spoke routing strategy for airlines in the deregulated environment. He demonstrates that hubbing has a significant effect on the cost structure in the extensive cost analysis he performs for the major carriers. The airlines with large hub-and-spoke networks have had two advantages over their competitors. They have realized certain cost economies pertaining to traffic density and network configuration about hubs, and they have been able to extract rents from passengers by being able to provide more connecting passengers with convenient on-line service than their competitors. In terms of the effects of network size and hubbing on profitability, the empirical evidence is inconclusive in this regard. The 'extent of hubbing' variable used in the analysis uses a ratio of the number of departures at each airline's 3% most utilized airports to the airline's total departures. In Chapter 3 of this book, these measures of hubbing will be compared with a new set of measures, and the main problems with the measures used by McShan will be set out there.

Jeng (1987, 1988) examines the impact of what he calls 'network parameters' (i.e. demand level, area size and number of nodes) on the form of the airline network system - specifically whether hub-and-spoke or point-to-point operations are better. He demonstrates that demand has a positive and very significant impact on point-to-point operations, but that in general, it is economical to incorporate the hub-and-spoke operation into the routing strategy (particularly, as the time value of schedule delay increases or enroute time/income decreases). Jeng uses the percentage of total enplanements for the top three airports served by each airline as a measure of hubbing.

Generally speaking, economists have made assessments of a wide number of aspects of commercial aviation since the passage of the ADA. In analyzing the airlines, these assessments have all been at the industry level, and would generally support the government's view that the passage of the Act was successful in improving competition in the industry, making air travel more economical, and thereby making it more widely available to the public.

Recently in the literature, some attention has been directed to the issue of network change and the role of the airline's system schedule in providing more economically efficient service. To date however, detailed and extensive testing of networks and their economic importance has not been undertaken. Conflicting results were reported in a number of different areas above and it was argued that the omission of a network variable was a contributing factor. Furthermore, two studies completely disagree on the importance of the network and the role of hubbing in airline operations, particularly airline productivity. One problem for economists attempting firm level, and indeed recent industry level studies, has been the availability of published data from government sources. Since the 1970s, CAB/FAA reporting requirements for certificated airlines have changed somewhat, so that for example, airlines no longer participate in the official 'origin-destination survey' of domestic and international traffic, nor do they need report the volume of connecting passengers on each of their routes. Much more data on air carriers' operations were available prior to 1970, and this made economic and other types of aviation studies easier. Detailed firm-level studies are difficult to undertake currently, again principally for data reasons. In such a highly competitive environment, it is not surprising that airlines are reluctant to open their operations to outside scrutiny. In this study, many of the questions posed are possible to address because Continental Airlines did make detailed route by route data available.

To begin to solve this controversy, we need to look at how those working in the industry view the airline's operation and where they place their emphasis in the management of these firms. The work done by operation researchers and schedule planners is reviewed in the next section.

2.2 Operations Researchers' and Transport Planners' View of Airline Operations

Given the central focus in this research on airline networks, and the conflicting views of the role and importance of the network in the economics literature, the literature on network design models was extensively reviewed so that the specific nature of the scheduling problem facing the firms would be grasped in detail, and also, the airline's problem could be set in the more general context of model structures in transportation planning and operations research. Papers examining issues such as hierarchical structures, dynamic models and spatial competition were covered also, so that again, progress and research directions in network construction for the airlines could be compared with research agendas in other areas. Little emphasis was placed on detailing solution algorithm procedures; the formulation of the problem and empirical studies using the models were the main areas of interest.

The generalized model of network design is defined for a set of nodes (discrete points located in a plane) and a set of arcs (links between the points) of the network. The variables in the model pertain to design decisions, the fixed cost elements, and to operating decisions, which are the variable cost elements. Each arc can be thought of as having a specific capacity, and the selection of this capacity depends on the cost decisions. The basic objective is to minimize total costs (fixed plus operating costs) subject to conservation of flow constraints and capacity constraints. There may be additional constraints depending on the problem in question. The form of the objective function will place greater emphasis on one or other of the cost variables, and as such the problem may be interpreted as placing greater emphasis on either efficiency throughout the network or accessibility over the network[4].

From this generalized model, minimal spanning tree and shortest path problems, facility location problems, travelling salesman (TSP) and vehicle routing problems (VRP), multicommodity flow problems, budget design problems, traffic equilibrium models and, fixed charge design problems may be derived. The model can consider directed or undirected arcs, for cases where the decision variables are discrete or continuous choice. For most of the problems mentioned above the aim is to assign flows so that the total routing cost is minimized. For traffic equilibrium problems, the users of the system exert greater control individually, and collectively may produce a routing policy that does not necessarily minimize the total routing cost.

Looking briefly at two of these problems, the TSP involves finding the minimum cost cycle through every node in the network. For the VRP, if each node is viewed as a customer to which a certain amount of a product must be delivered from a central depot, then with 'n' vehicles each having a maximum capacity of 'k' units, the problem is to route the vehicles so that all the demands are satisfied (nodes) and the total distance is minimized (flows), remembering that all vehicles are required to return to the depot. The VRP reduces to a TSP when dealing with only one

[4]Bunge (1966) demonstrated six possible solutions to the network problem for a system with five nodes; the solutions were obviously different when the objectives were, for example, the 'minimum cost to build' system and the 'minimum cost to user' system.

vehicle of infinite capacity. The basic conclusion that may be drawn from the academic literature on VRPs is that exact solutions to these problems are cumbersome for small-medium sized models and computationally impossible for large scale problems [Magnanti and Wong (1984), Wong (1984)]. These VRPs in many cases form the basis for airline scheduling models.

For an airline, the optimization of the flight schedule (i.e. construction of a schedule) is the most critical step in the firm's operation, since practitioners argue that it is the flight schedule which defines not only the product of the firm, but also to a large extent the production plan. The construction of the flight schedule is central to finding the most effective and efficient deployment of an airline's resources, and if the market and regulatory mechanisms work properly, to obtaining an air transportation system which satisfies public objectives [Etschmaier and Mathaisel (1984, 1985)]. Typically, the objective for the airline is to find a set of routes and frequency of service, aircraft and departure/arrival time assignments which maximize profits under fixed competitive and economic conditions. The airlines make these decisions based on information they possess (or estimate) on each origin-destination pair's (each market's) demand and revenue functions, over the time of day and weekly planning cycles. They also input information on distance, time and operating restrictions on each route, aircraft operating costs and other characteristics, and operational and managerial constraints.

It is claimed that the airline's objectives cannot be expressed purely in monetary terms, so that even if it were possible to solve such gigantic combinatorial problems, additional complications are ever present. Many constraints defy inclusion in the model given their changeable nature and indeed many details of the schedule environment are constantly in flux; consequently there often arises the need for major adaptations of the optimization algorithm [AGIFORS (various years)].

For other modes, it is possible to break down the scheduling problem into subproblems, but airlines need to solve the whole scheduling problem since the number of aircraft and frequency on routes, for example, are too small to permit rounding, and combinations and permutations are too large to be done by hand. Additionally, the need to serve several markets often makes it impossible to break down the network. There is a large number of feasible alternate utilizations of the fleet; crew scheduling is important and costly; the number of route regulations and maintenance requirements is large and finally, the requirements in terms of service facilities at stations are fairly extensive [Etschmaier and Mathaisel (1984, 1985)].

In assessing a schedule, opposition to the concept of separate link by link or route by route evaluation has been expressed [Spry (1984)]. This author suggests that given the joint product nature of air transportation, the cost of joint product activities has to be largely a matter of judgment, convention and allocation. The airline's route network evolves from a variety of commercial, political and social decisions. As such, the economics of each of the individual routes that go to make up that network may only make financial and commercial sense when related to one another in an integrated way. What we learn from this literature on airline scheduling is that the airline's operation and organization is very different to other industries and even to other transport industries. Those working in the area very much highlight the importance of the network as essentially the bones of the operation upon which all else rests.

Any model intended to optimize an airline's flight schedule has never been used on a routine basis, and even the use of a sequence of models involves such rigidities as to render the system useless [Etschmaier and Mathaisel (1984, 1985)]. An interactive system, where human and computer each perform those tasks to which they are best suited, has been the focus for most airlines operating reasonably large systems [AGIFORS, (various years)].

In constructing a schedule, heuristic procedures are sometimes utilized to compose the schedule on a flight by flight basis. Alternatively, some airlines use a stepwise approach, where the construction process is broken down into a 'frequency planning subproblem' (selection of routes and determination of frequency of service on each route), a 'time of day' subproblem (where departure times are determined, based on demand, connections etc.)

and, an aircraft rotation subproblem, where departure times are checked for operational feasibility. The choice as to which approach to take depends on such things as the airline's structural characteristics, route structure and market structure. Other approaches consider the fact that in a given year, a schedule will to a greater or lesser extent be similar to the schedule for the preceding period, and the problem of schedule construction could be viewed more as a problem of finding the best modifications to the old schedule. Better organization of the existing network is then an improvement in technology since it represents a more efficient production plan. By continually iterating between schedule construction and schedule evaluation, the airline's flight schedule is obtained. While operations research methods cannot solve these problems *per se*, they have made a major contribution in terms of the conceptualization and formulation of the scheduling problem.

The literature on scheduling indicates the combinatorial difficulties in solving the routing problem and the complexity of even attempting to capture aspects of an environment continually in flux. For the economic analyst, it has to be clear that the possibility of technical or production inefficiency is high, given that optimal solutions cannot be found, and that so many continual readjustments need to be made all of the time in the firm. The economist working at the firm level can make a contribution by pointing out the relative inefficiency of one route vis-à-vis the rest of the system, once an effective method for accounting for the 'interconnectedness' of the system has been derived.

2.3 Geographical Aspects to U.S. Air Traffic Distribution

In the 1950s, geographical research in airline transportation focused principally on the relationship between the U.S. urban hierarchy and air traffic hierarchies. While traditional economic geography studies of this kind have not been continued in the 1970s and 1980s, it is speculated that one impact deregulation has had is to weaken the relationship between these two hierarchies, certainly at the intermediate level, since traffic concentration at many regional hubs bears little if any resemblance to the size of the hinterland population. These notions will be explored in Chapter 3.

Taaffee (1959), in examining the relationship between the size of the U.S. urban hierarchy and air traffic, notes that at least in the 1950s, air transportation was concentrated in the medium- and long-haul carriage of passengers; he considers the effects of urban characteristics such as size, function, proximity to other cities and railroad services on air traffic, concluding that urban population and grouping have been the two primary factors in the development of the air passenger pattern of the U.S.. Secondary factors like urban function and overnight rail service also played a role, as did poor surface connections, strategic route situation and airline policy. He notes that prediction of changes in U.S. air passenger patterns is essentially an attempt to predict the direction of technological development, and as such, a worthwhile exercise.

In a case study, Taaffe (1956) focuses on a series of maps depicting trends in Chicago's air passenger traffic during what he calls the critical growth period of 1949-55. He attempts to demonstrate a relation between growth rates and possible growth-promoting characteristics of individual routes, highlighting the effects of recreation travel, length of haul and low coach fares. In a later paper Taaffe (1962) attempts to compare the hierarchy formed by air passenger linkages between U.S. cities, with gravity model expectations. The passenger air linkages represent an empirical cross section from the upper levels of the urban hierarchy in the U.S.. He demonstrates the increasing dominance of the large centers (New York, Chicago, Los Angeles and San Francisco), the role and importance of particular regional ties, and the likely intensification of the hierarchical nature of inter-city linkages.

Recently, Kanafani and Ghobrial (1984) have examined the implications of airline hubbing for airport economics in the U.S.. They point out that, at a number of airports, connecting traffic resulting from airlines' propensity to 'hub', dominates the airport's traffic flow, and accordingly, changes fundamentally the economic role of the airport in the region it serves. With these high connection rates, passenger flow becomes primarily a matter of processing and movement from one gate to another. The authors argue that the revenue generation potential at these airports has decreased, and in general, on a per-passenger basis, the airport does not generate the same amount of revenue as it would for originating traffic, even though the fees received for use of the airport facilities are the same. For these hub airports, the dependence on airline networking is considerable, when compared with dependence on the economic demand for travel within the region it serves, and in addition, a predominantly 'connecting-hub airport' does not contribute as much to its local economy as one that serves mostly originating traffic. While the authors seem to highlight the disadvantages of hubbing from the local community's point of view, the advantages of hubbing may be more forceful from the community's and the nation's perspective. Airports do not exist in isolation and an airport's efficiency and contribution cannot be examined purely in terms of local effects. The greater frequency of flights, the greater range of routes and reduced operating costs (resulting in many cases in fare reductions for consumers) have benefited communities across the U.S. since 1978. This is demonstrated in Chapter 3 in the increased volume of traffic annually, and in the expanded size of domestic carriers' networks.

Looking now at models of traffic potential, the family of gravity models calibrate the relationships for interaction between pairs of nodes in a system, using variables like their population size and separating distance. Using the basic gravity model relationships, Fotheringham (1983) derived what he calls a 'competing destination' model, and using 1970 CAB data, the author computes 'origin and destination accessibility parameters' for the leading 120 hubs in the U.S. air system in that year. In the usual gravity models, the ratios of predicted interactions between any two pairs of points are independent of the distribution of destinations, simply because the distance-decay parameters estimated are independent of the spatial structure. In the model Fotheringham proposes, this bias in the distance-decay parameter is removed and origin or destination specific assessments of accessibility to all other nodes in the spatial system may be made. The author provides a series of maps showing the spatial variations in these parameters and concludes that in general, gravity models overpredict interactions at most of the accessible nodes in the north eastern U.S., while they underpredict interactions with the inaccessible nodes in the rest of the country. Since these traditional models do not take competition among destinations into account, Fotheringham suggests that a relationship between destination accessibility and interaction be included to reduce errors.

Fotheringham's research is one of the only recent studies of airport and urban hierarchy links, though his test data set is from 1970. Because of the changes in CAB reporting requirements, an updated study calibrating this competing destinations model for the U.S. in the post-deregulation period cannot be undertaken. This kind of analysis has great potential for airport managers, airlines and state and federal government policy makers. For aviation analysts, it offers a macrolevel perspective on the spatial reorganization of air traffic flows and their impact on the national network of airports that took place after 1978.

In the economics literature on airline deregulation reported in this chapter, a wide range of issues were included, and, in many instances there were disparities in the results and conclusions. When examining market structure for instance, economists are unable to confirm the contestability hypothesis for subsets of markets; they look at single markets (routes) and make their comparisons at this level, not considering the joint product nature of the airline's production process nor different network organization schemes employed by the airlines. A better understanding of the network structure and the role of network organization in the production process should help improve the specification of models testing market structure. At the industry level, an accurate measure of each firm's network structure is needed to properly account for network organization and to standardize the precise meaning and role of networks. By consistently using such a measure, the possibility of contradictory results is reduced.

The microlevel focus in the schedule construction and evaluation procedures provides a very useful perspective on the role and characterization of the network from the point of view of the individual firm. By generalizing the key components of the network at this microscale, the essential characteristics of routes comprising the network may be incorporated into an economic model of the individual firm.

At the national level, this chapter highlighted the paucity of research focusing on national air traffic patterns in the last 20 years. Geographical research in the 1950s and 1960s examined the relationship between the urban hierarchy and the air traffic hierarchy, but no recent research has been undertaken in this area. The impact of carrier network changes since deregulation on the nation's air traffic flows needs to be quantified, and this requires that both airport and airline traffic distributions can be described quantitatively, in such a manner as to be able to trace change over the last two decades.

CHAPTER 3
THE SETTING: TRENDS IN U.S. COMMERCIAL AVIATION

The purpose of this chapter is to survey the changes that have been taking place in the airline industry and in the national airways system, since the 1960s. In the previous chapter, various economic assessments of the industry since 1978 were reported, and two consistent shortfalls in the examinations and analyses were highlighted. These were the failure to focus on the extent and nature of specific network changes, and the effect of these changes on the national airport system. In this chapter, both of these issues will be addressed. Specifically, the aims are to (1) describe the changes taking place in the national airways system and attempt to link these with patterns of change among air carriers' route systems; (2) to ascertain whether the air carriers' propensity to concentrate traffic in their system at a small number of points (i.e. to develop and emphasize so-called 'hub-and-spoke' systems) has occurred only since 1978, or if this trend was in place before deregulation; (3) investigate the most appropriate measure of networks and network change, taking into account as many aspects of the airlines' networks as possible.

In the next section, a description of the national airways system is given, and the pressing spatial questions are emphasized. The subsequent sections describe the airport system, the traffic distributions over the system in the last 20 years, and the supporting air carrier industry. These carriers are then examined closely; industry characteristics are set out, and two airlines' traffic patterns are carefully analyzed for a 20 year period. The Gini index of concentration is used extensively in the following sections, and justification for its use is provided.

3.1 The National Airways System

The provisions of the 1926 Air Commerce Act, and later the Civil Aeronautics Act of 1938, made clear the federal government's commitment and interest in a national airport system, and in airport development. Under these bills, the federal government assumed responsibility for the development and operation of the airways and for the control of air traffic. A CAA report of March 1939 pointed out that an airport's usefulness depended on its location relative to other airports in the nation, and though communities may rival one another in the attractions that their respective airports offer, their common interest is far more important than their rivalry [CAA (1939)]. They concluded that all users and communities have good reason for maintaining an adequate system of airports serving every part of the nation.

The FAA, under the Department of Transportation control the airway system, and categorize airports and airfields into two groups: (1) those airports used for air carrier operations which serve the needs of scheduled air carriers and (2) those for general aviation, serving all other civil air carriers (air taxis, business, agriculture etc.). In this book it is the former category only which is the focus of attention. For this category, the FAA developed its air traffic hub structure, and they use this classification scheme to measure the concentration of all civil air traffic in their economic and operations research procedures [CAB (1985)]. The FAA point out that social, economic and other factors influencing a community's ability to generate air carrier traffic are brought together in this one medium. The FAA air traffic hubs are not airports *per se*. Rather, they are the cities and Standard Metropolitan Statistical Areas (SMSAs) requiring aviation services. These communities fall into a four hub classification scheme, depending on the community's percentage of the total enplaned passengers in all services and operations of U.S. certificated route air carriers within the 50 States and other designated areas. The hub classes are given below in Table 3.1.

Airline 'hubs', in comparison, usually refer to those key airports (typically two or three) around which the entire route system is organized. Airline hubs emphasize the transfer of passengers, and accordingly, at these airports, the airlines' offer a large number of connecting flights to other areas. The airlines construct what they call 'complexes', which are large blocks of flights arriving at the hub airport within a very short period of time, and this is followed by an equally large number of departures, when passengers have made their connections. The hub airports usually have between 3 and 5 complexes in operation every day. The term 'hub' in the context of airline schedules is more loosely applied.

Table 3.1
FAA Hub Classification Scheme

FAA hub type	Percentage of total US passenger enplanements
Large	1.00% or more
Medium	0.25 to 1.00%
Small	0.05 to 0.25%
Nonhub	less than 0.05%

The FAA operate and fund the National Airway System, a broad planning framework where aviation demands and activities of metropolitan areas, states, regions or indeed the nation may be outlined with solutions suggested for a sequence of action to meet the facility demands. Individual airport master plans detail specific design recommendations for the facility.

While most large cities began studies of new airports 20 or more years ago, very little progress has been made in the 1980s in terms of designing and building new large airports. Since the completion of facilities at Dallas-Fort Worth airport over a decade ago, no major airport has been built in the U.S., though plans were recently outlined for a new 'superhub' at Denver, Colorado. Investment at the major airports in the recent period has been aimed more at upgrading facilities rather than expanding capacity.

At the lower end of the hierarchy, carrier network changes could involve a small community either losing all of its service, or experiencing phenomenal growth in terms of demand for its facilities, but not being able to satisfy this demand. Given the volatile nature of the market and traffic patterns at smaller airports in particular, the investment and planning process aimed at maintaining and structuring the National Airway System has been made considerably more risky. Some major carriers have been increasing their traffic feed to larger hubs by aligning themselves with small carriers, or by directly serving the smaller communities. The airlines have tended to select underused airports in expanding their systems, since these airports would be comparatively free of constraints that restrict many of the larger airport facilities, such as airside and groundside capacity, environmental concerns and airport policies that inhibit further growth. For the FAA, long term planning goals are difficult to formulate when projections of traffic need to be revised as carriers continue experimenting with new routes and connections, but fail to provide advance information to the FAA on their expansion or contraction plans.

For the smaller airports being selected by the airlines, the grounds on which they may deny access are limited and difficult to establish now under deregulation. It is up to the community to make a case to the FAA, who are more likely to side with the airline. Deregulation allowed airlines to determine their own routes, rates and service frequency, but no mechanism was left for determining which airlines should get access when an airport has limited facilities [Quinones (1986)].

Because of contraction in service at many local airports, the real estate value of the airport may suggest other, more efficient uses for that land. However, if such an airport has received any funding from the FAA for facility improvements, part of the requirement for funding was that the airport remain open for the succeeding 20 years. In 1987, it was estimated that 2350 airports and airfields fell into this category nationally [Weiner (1987)].

The structuring of the air carrier's route system has wide ranging implications not just for the firm's profitability and competitive standing, but also for the nation's airway system. The stability and sustainability of this system depends on the air carriers' choice of airports, their service frequency and the possible connections that may be generated. The understanding of these relationships is important for federal agencies in order for future investments and planning to be effective and efficient.

3.2 Patterns in U.S. Commercial Aviation 1969-88

Since the early 1960s, U.S. passenger traffic has been growing steadily at an average annual rate of 5.8%, with minor declines in 1973/74 and 1980/81 due to the impact of the oil crises. In the case of the 1980/81 drop, decline must also be attributed to a correction after unusually high enplanement figures for 1978 and 1979, the deregulation year. Figure 3.1 shows the traffic trend for the period 1969-88, where traffic is indexed to 1977. International traffic has typically accounted for between 3 and 5 percent of the total U.S. traffic over this period.

The increase in air traffic has not been equally distributed over the U.S. airport system. Some airports' traffic levels have grown at rates greatly exceeding the national average while others have declined and a substantial proportion have ceased to serve the scheduled traffic market. The changing distribution of traffic can be very simply captured by computing rank correlation coefficients of the top 100 airport's ranks in 1987, with their ranks in 1985, 1981, 1977 and 1973. The vast majority of scheduled air traffic takes place at these top 100 'hubs'. Table 3.2 shows the percentage of U.S. traffic at these hubs in the four time periods mentioned above, and the proportion of all airports/communities certificated to receive scheduled service that the top 100 represent.

Table 3.2

Proportion of traffic at top 100 U.S. hubs 1969-88, and correlation coefficients of the top 100 hubs' ranks in 1987 with other years.

Year	% Traffic	Proportion of CPs[1]	Total no. of CPs	Correlation with 1987 ranks
1969	96.9	27.9	524	-[2/]
1973	92.5	13.0	772	0.707
1977	92.3	15.7	638	0.749
1981	94.7	15.9	627	0.848
1985	96.0	26.7	374	0.953
1987	97.7	26.2	477	1.000
1988	97.7	27.0	463	-[3/]

Source: CAB/FAA Airport Activity Statistics of the Certificated Route Air carriers, Washington D.C..

1/ CPs refers to certificated points.

2/ The correlation coefficient was not computed for 1969, since it was not possible to determine the ranks of all hubs in that year.

3/ Not computed for 1988

U.S. PASSENGER ENPLANEMENTS
Traffic Index 1969–88

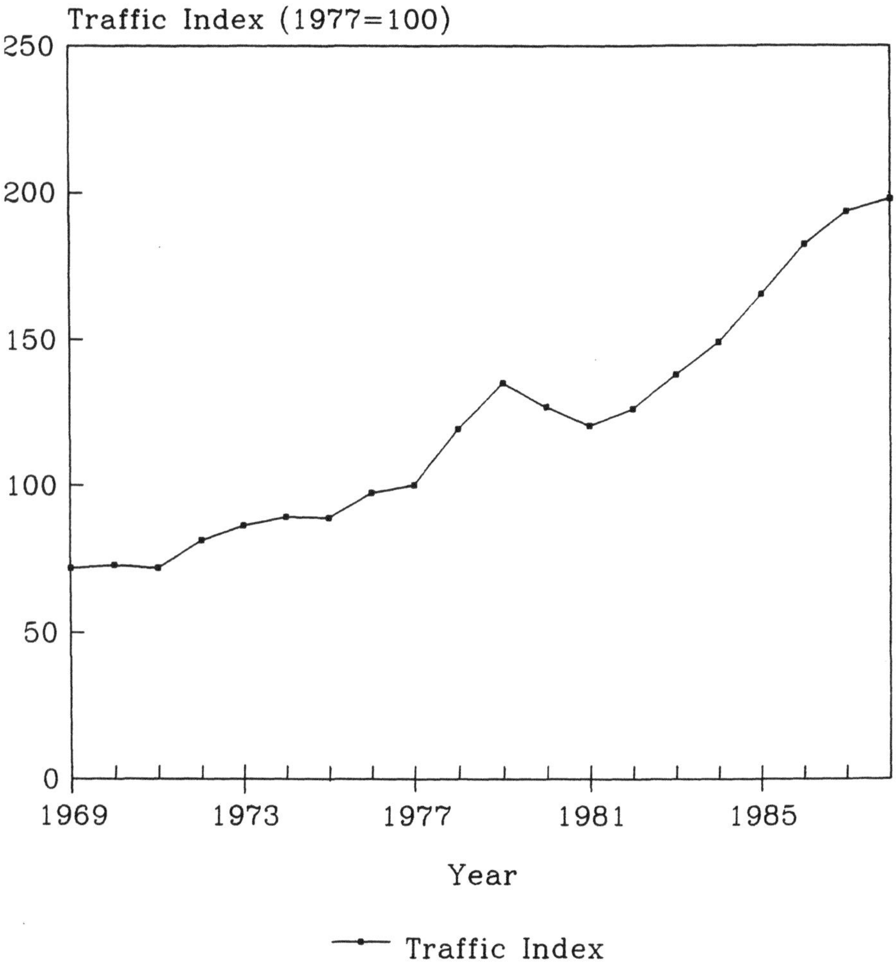

Source: CAB/FAA Data

Figure 3.1

The total number of FAA certificated points has dropped substantially since deregulation, and so a smaller number of airports/communities are handling this greatly increased volume of traffic in the 1980s. Figure 3.2 shows the total numbers of FAA certificated points and, the total numbers of large, medium and small hubs for the period 1969-88. In Table 3.3, the traffic proportions at all large, medium and small hubs for the period 1969-88 are summarized. The number of large and medium hubs in the system has remained relatively constant in the last 12 years, though the number of small hubs has fallen noticeably, and accordingly so has the proportion of traffic handled at these nodes. U.S. air traffic has become considerably more concentrated at a smaller number of centers. It is most likely that many of the smaller communities began receiving alternative service when scheduled service ceased, but these data are not as easily or readily available.

Table 3.3
Traffic at large, medium and small FAA hubs 1969-88

Year	Large hubs		Medium hubs		Small hubs	
	No.1/	% traffic2	No.1/	% traffic2	No.1/	% traffic 2/
1969	22	69.7	38	20.4	86	9.9
1973	25	68.4	39	18.4	84	9.0
1977	25	68.1	39	18.4	92	10.2
1981	24	70.2	40	19.0	72	7.9
1985	26	72.8	37	18.1	61	6.7
1987	28	74.1	34	16.6	63	7.1
1988	29	74.6	32	15.9	64	7.3

Source: CAB/FAA Data.
1/ Number of large, medium and small hubs (note: hubs refer to SMSAs and communities, not airports, so that there may be more than one airport at the FAA hub)
2/ % traffic refers to the proportion of U.S. domestic passenger enplanements at large, medium and small hubs.

Along with the reduction in the number of centers receiving scheduled air services, the spacing of the remaining airports has also changed over the recent decades. In Figure 3.3, the average distances between large hubs and other large, medium and small hubs for the period 1965-85 are presented. Since the 1960s, it can be observed that the mean distance separating large hubs has been

U.S. FAA HUBS 1969–88

Number of large, medium and small hubs, and certificated points (CPs)

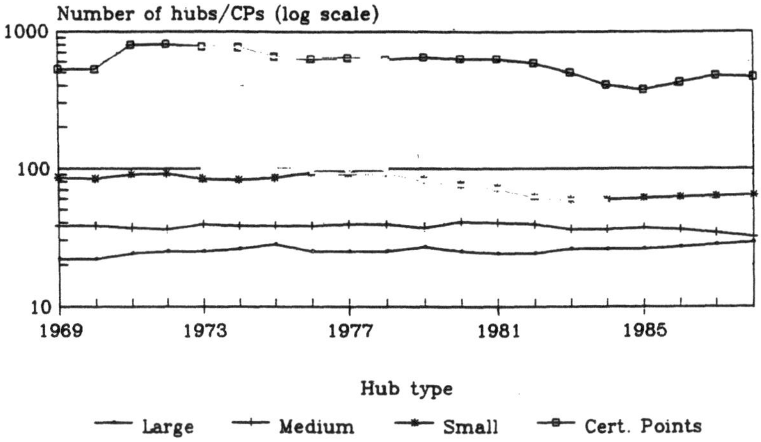

Figure 3.2

HUB SPACING 1965–85

Mean distance between hub types

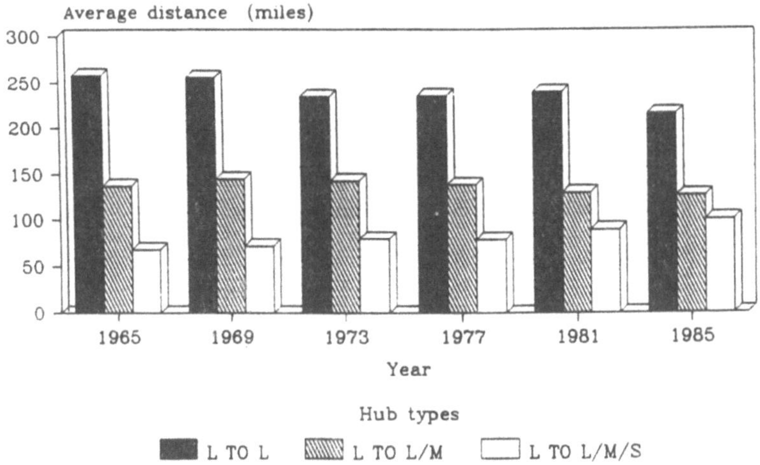

Figure 3.3

reducing gradually. The most stable trend is noted for the large to large or medium hub figures. The average mileage between large and all other hubs has been increasing over the period, though the increase is more substantial after 1977. The trends reflect the contraction of the national airway system and the more efficient spacing of the remaining airports.

3.3 Measures for Summarizing Airline and Airport Traffic Distributions

In the previous section, the tremendous increase in passenger enplanements was traced for the U.S. over a 20 year period. In this section, the traffic distributions at a sample of those airports are examined so that the pre-1978 patterns of service by carriers may be compared and contrasted with the patterns observed in the deregulated years. In the government and academic literature on deregulation, many authors often refer to the noticeable change in carriers' routing strategy from predominantly linear, or point-to-point networks, prior to 1978, to 'hub-and-spoke' structures in the 1980s. Most of the evidence for this comes from visual inspection of route schedules and from examining traffic proportions at carriers' busiest airports. The aim in this section is to present data for a longer time period (i.e. 1969-88), and measure this propensity by the airlines to use hub-and-spoke routing systems, and in addition, to examine the effects of this hubbing phenomenon on the U.S. airport system.

It was noted in Chapter 2, that some attempts were being made by economists to include a measure of airlines' network organization in their analysis. In the two major studies focusing on the role of networks, both authors used the percentage of traffic at the first, second, third and fourth busiest airports (or the top 3% airports) for each airline, with McShan also computing the Herfindahl index in his study [McShan (1986), Kanafani and Hansen (1985)]. Caves et al. (1984) merely count the number of airports served and as mentioned before this does little if anything to address the issue of the airline's organization of traffic through the nodes they serve.

More generally in the economics literature, considerable

attention has been given to the selection of appropriate measures of concentration or dispersion for various types of distributions, perhaps most noticeably, the distribution of income [Atkinson (1970, 1975), Gastwirth (1972), Kuznets (1955), MacDonald (1984), Paglin (1975), Shalit (1985), Sen (1976)]. While measures like the mean and variance of the distribution of income have been used in the past, most writers would agree that such measures as the Gini index of concentration, Theil's relative entropy measure and the coefficient of variation are far superior statistics, since they yield unit free comparisons that are statistically sound and make more sense intuitively and theoretically. It is surprising that in selecting measures for summarizing air traffic distributions, more attention has not been given to the extensive literature on summary measures in other branches of economics.

Those measures used in the economics airline studies mentioned above will be contrasted with the Gini index, the coefficient of variation and Theil's entropy measure. Briefly introducing these statistics, the Herfindahl index [Herfindahl (1950)] is computed as

$$H = \sum_i (x_i/\Sigma\, x_i)^2$$

where x_i is the air traffic at airport i (or in the case of an airport's traffic distribution, x_i is the air traffic handled by carrier i)[5]. The term 'air traffic' refers to passenger enplanements, unless otherwise indicated. The main problem with this index lies in the fact that it is size dependent and some argue that it is a poor measure since it can be derived from the parameters of the underlying size distribution [Gaile (1984)]. It will be shown later in this chapter that the index is relatively insensitive to changes in the number of airports served (in the airlines' case), and to the redistribution of traffic among these airports over different time periods.

The Gini index of concentration is defined as

[5]The index is commonly used to measure industrial concentration. Indeed in Antitrust cases, this index is the preferred measure used by investigators to gauge the degree of concentration in particular economic markets, including air traffic markets [General Accounting Office (1988)].

$$G = \frac{1}{2n^2}\,\bar{y} \sum_i \sum_j | y_i - y_j |$$

where the ys (air traffic at i or j) are ranked in increasing order, and $\bar{y}$ is the mean. This measure is theoretically sound and satisfies many intuitive properties which measures of concentration should have[6].

In the analysis of income distribution, the poverty line is a pre-defined level of income, and income inequality is measured against that income level. In the airport and airline analysis, the equivalent line measures the median traffic; increases in traffic at points below the median reduce the level of concentration while increases in traffic at points above the line increase the concentration level. One of the biggest single problems with the Gini index lies in the fact that it is difficult to compute when grouped data are being used. However when single continuous variables are available, Shalit (1985) has demonstrated that it is easily computable and he endorses the measure as a sound and useful one in a wide variety of contexts. The Gini Coefficient can be shown graphically when 'Lorenz curves' (showing the observed distribution of a variable) are plotted.

The coefficient of variation is computed by dividing the arithmetic mean of a distribution by the standard deviation. Taylor (1977) recommends caution when using this measure for

6 Sen (1976) examined the properties of income inequality measures, and proposed his monotonicity, transfer, relative equity and ordinal rank weight axioms. Essentially, these axioms suggest some intuitive properties that a good inequality index should possess. He proposed the Gini index as the only measure to satisfy all of the axioms. In terms of airline and airport traffic distributions, these axioms require that
(1) a concentration measure is increased if a lower ranked airport's traffic proportion is reduced (monotonicity axiom)
(2) a pure transfer of traffic from a low ranked airport to a high ranked airport will increase the concentration measure (transfer axiom)
(3) if airport i is accepted to have a smaller traffic quotient than airport j in a given distribution, then the weight on the traffic shortfall for airport i should be greater than the weight of the traffic shortfall j (relative equity axiom)
(4) the weight on the traffic gap of airport i equals the number of airports in the network with at least the same traffic proportion as airport i (ordinal rank weight axiom).

comparative purposes, since the variation measured depends on the number of cases in the distribution.

Finally, Theil's entropy measure is computed generally as

$$T = \sum [(x_i / \Sigma x_i) \log (x_i / \Sigma x_i)] / \log n$$

where x_i is again air traffic at airport i. This is just one of a family of entropy measures which include various set decompositions allowing for within and between variations of the sample distribution to be measured and compared. The measure is popular in the literature, since it is theoretically and economically sound.

Simple measures such as the range of a distribution and those proposed by McShan and others are poor indicators since they take little or no account of the number of airports served, they are scale dependent, and they are insensitive to all variations other than those at the extremes. Accordingly, their utility and appropriateness are questionable and below a more thorough demonstration of these facts will be given.

A random sample of five large, five medium and five small FAA hubs were selected from the 1987 Airport Activity Statistics of the Certificated Route Air Carriers. The total domestic scheduled and nonscheduled passenger enplanements were recorded for each certificated carrier serving the hub, and these traffic distributions were detailed for the years 1969, 1973, 1977, 1981, 1985 and 1987, (five year periods, with 1977 being used as a base year, since it was the last fully regulated year; 1987 was the most recent year for which data were available, and is included for this reason).

Tables 3.4, 3.5, and 3.6 record the number of carriers, the percentage of U.S traffic, the Gini index and the coefficient of variation for each hub, for each of the 6 years. For the five large hubs (serving from between 1.65% and 5.30% of total U.S. traffic), the average number of carriers increased from 8.2 in 1969 to 15.2 in 1987. This increase in the number of carriers is quite marked after 1977, since the ADA allowed carriers to enter new markets freely after 1979. There was an average of 17 carriers serving the large hubs in 1985, though this number dropped in 1987, due most likely to the contraction in the total number of carriers serving the industry, because of mergers,

Table 3.4
Characteristics of sample FAA large hubs, 1969-87

Airport	Year	Gini Index	Number of Carriers[1/]	Traffic Proportion[2]	Coefficient of variation
Atlanta, GA.	1969	0.6972	9	5.04	0.6592
	1973	0.7049	9	5.98	0.6469
	1977	0.6806	8	6.37	0.6819
	1981	0.6470	9	7.03	0.7691
	1985	0.8116	13	5.69	0.4566
	1987	0.8193	13	5.31	0.4511
Denver, CO	1969	0.5574	10	2.21	0.9387
	1973	0.5584	10	2.59	0.9372
	1977	0.5876	11	2.97	0.8674
	1981	0.6672	16	3.92	0.6768
	1985	0.7304	18	3.82	0.5406
	1987	0.7526	13	3.65	0.5067
Las Vegas, NV	1969	0.3455	7	1.11	1.5764
	1973	0.3522	7	1.29	1.5620
	1977	0.3164	7	1.55	1.7071
	1981	0.5319	18	1.59	0.9110
	1985	0.5039	24	1.27	1.0689
	1987	0.5518	20	1.60	0.8288
Pittsburgh, PA	1969	0.4345	7	1.95	1.2302
	1973	0.5434	7	1.86	0.9472
	1977	0.4482	6	1.82	1.1706
	1981	0.6991	10	1.57	0.5766
	1985	0.7747	11	1.93	0.4061
	1987	0.7911	11	1.91	0.3847
Seattle, WA	1969	0.5948	8	1.44	0.7780
	1973	0.4889	9	1.35	1.0344
	1977	0.4992	10	1.55	1.0418
	1981	0.5963	18	1.61	0.7901
	1985	0.5489	19	1.58	0.8952
	1987	0.5905	19	1.60	0.8144

Source: CAB/FAA Data.
1/ Number of certificated carriers offering scheduled and non-scheduled passenger services
2/ Proportion of passenger enplanements at hub airports compared with total U.S. enplanements

Table 3.5
Characteristics of sample FAA medium hubs

Airport	Year	Gini Index	Number of Carriers[1]	Traffic Proportion[2]	Coefficient of variation
Austin, TX	1969	0.2736	3	0.17	1.6707
	1973	0.1883	3	0.17	2.5000
	1977	0.2268	3	0.21	1.9942
	1981	0.4701	8	0.36	1.0566
	1985	0.5910	12	0.50	0.8098
	1987	0.6075	12	0.45	0.7957
Indianapolis IN	1969	0.3953	6	0.63	1.3924
	1973	0.3807	7	0.63	1.4466
	1977	0.2660	6	0.60	2.0421
	1981	0.4597	10	0.47	1.2201
	1985	0.5168	14	0.48	1.0136
	1987	0.5313	15	0.53	0.8897
Columbus, OH	1969	0.4293	7	0.53	1.0186
	1973	0.4092	8	0.50	1.2943
	1977	0.5033	9	0.48	1.0021
	1981	0.5103	10	0.42	0.9367
	1985	0.3498	10	0.42	1.6052
	1987	0.4896	14	0.40	1.1162
Raleigh-Durham, NC	1969	0.4165	3	0.30	1.1313
	1973	0.4353	4	0.31	1.0363
	1977	0.5050	4	0.29	0.8820
	1981	0.5470	7	0.31	0.9235
	1985	0.5786	11	0.37	0.8227
	1987	0.5814	10	0.54	0.8329
Nashville, TN	1969	0.5221	9	0.42	0.8639
	1973	0.4673	9	0.40	1.0500
	1977	0.3019	8	0.42	1.7724
	1981	0.2974	9	0.38	1.8588
	1985	0.3992	12	0.38	1.3913
	1987	0.6262	11	0.70	0.5685

Source: CAB/FAA Data

1/ Number of certificated carriers offering scheduled and non-scheduled passenger services

2/ Proportion of passenger enplanements at hub airports compared with total U.S. enplanements

Table 3.6
Characteristics of sample FAA small hubs

Airport	Year	Gini Index	Number of Carriers[1]	Traffic Proportion[2]	Coefficient of variation
Little Rock, AK	1969	0.2936	5	0.20	1.5413
	1973	0.3697	5	0.19	1.3038
	1977	0.4057	5	0.20	1.1692
	1981	0.3920	5	0.17	1.1182
	1985	0.2141	6	0.20	1.9565
	1987	0.2317	6	0.21	2.0390
Daytona Beach, FL	1969	0.2366	2	0.08	2.1135
	1973	0.3219	2	0.11	1.5532
	1977	0.3719	2	0.12	1.3445
	1981	0.4879	4	0.10	0.8904
	1985	0.2461	3	0.06	1.9515
	1987	0.3454	6	0.09	1.1958
Portland, ME	1969	0.0000	1	0.05	0.0000
	1973	0.0000	1	0.07	0.0000
	1977	0.3738	2	0.08	1.3376
	1981	0.0000	1	0.08	0.0000
	1985	0.2367	3	0.13	1.7758
	1987	0.2202	5	0.13	1.9954
Harrisburg, PA	1969	0.2004	2	0.13	2.4951
	1973	0.3633	2	0.12	1.3764
	1977	0.2398	2	0.12	2.0854
	1981	0.1149	2	0.07	4.3511
	1985	0.2380	4	0.08	1.7715
	1987	0.3578	7	0.09	0.9988
Brownsville TX	1969	0.1107	2	0.04	4.5186
	1973	0.4903	2	0.00	1.0197
	1977	0.1858	2	0.00	2.6905
	1981	0.5289	4	0.14	0.7986
	1985	0.3436	4	0.12	1.1979
	1987	0.3443	4	0.11	1.2314

Source: CAB/FAA Data

1/ Number of certificated carriers offering scheduled and non-scheduled passenger services

2/ Proportion of passenger enplanements at hub airports compared with total U.S. enplanements

takeovers/acquisitions or less typically, financial failure. One of each hub type was selected at random, and traffic distributions at two of these hubs for three different years are illustrated in Figures 3.4 and 3.5. These figures show the traffic handled by air carriers serving the airport; the air carriers are ranked for each airport depending on the proportion of traffic that they handle. The 45-degree line represents the distribution of traffic if all carriers handled the same proportion. The Gini index measures the ratio of the area between the 45-degree line and the actual traffic distribution, to the area of the triangle enclosed by the 45-degree line and the two axes.

The Gini index, measuring the difference between actual traffic distribution and a traffic distribution where all carriers carried the same proportion of traffic, rose from an average of 0.526 in 1969 for large hubs, to an average of 0.701 in 1987. A Gini index of 0.5 is quite high under any circumstances, so it must be stressed that almost a decade before deregulation, traffic patterns at large U.S. hubs exhibited a high degree of concentration, and after passage of the Act, this tendency became ever more marked. The values of the Gini index at the 15 hubs for each of the 6 years were regressed on the number of carriers at each hub, and on the proportion of traffic at each hub.

The Gini index is positively and significantly correlated with the number of carriers serving each airport (the R-squared is 0.46). The Gini index is also positively and significantly correlated with the proportion of traffic at each airport (R^2 = 0.44). Regressing the Gini index on both of these variables gives an R^2 of 0.63. The relationship between the proportion of traffic and the number of carriers is very weak, the R^2 value being 0.19. Thus the busiest airports do not necessarily have a large number of carriers serving them, but these airports are likely to have a more concentrated traffic distribution in comparison to the smaller airports. Those airports being served by a larger number of carriers are also likely to have a more concentrated traffic distribution.

The dramatic increase in concentration, particularly at the largest hubs, reflects the large airlines' consolidation of their "hubs" at these airports in the deregulated environment. The

CONCENTRATION AT FAA HUBS

Traffic distribution at a sample large hub: Las Vegas, NV.

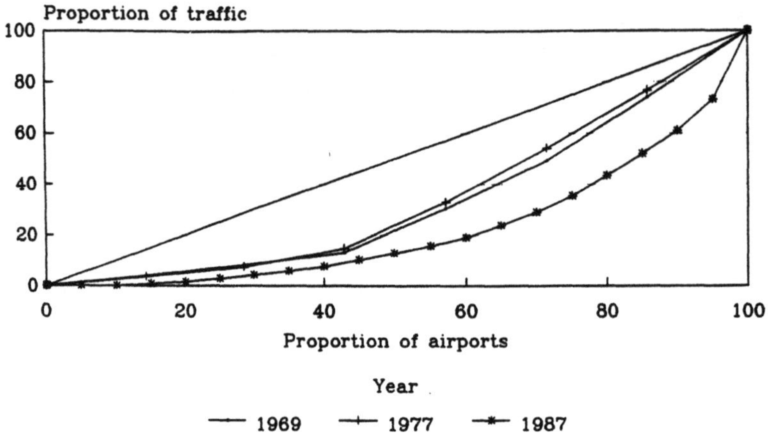

Figure 3.4

CONCENTRATION AT FAA HUBS

Traffic distribution at a sample small hub: Portland, ME.

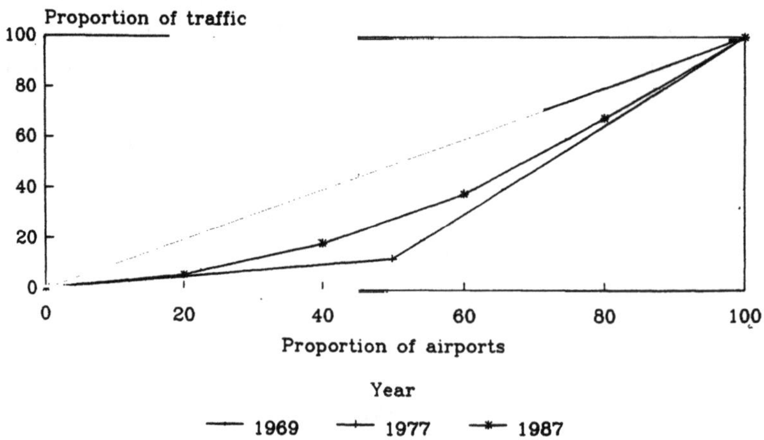

Figure 3.5

concentrated nature of the traffic distribution prior to 1978, indicates that many carriers had already selected and grown around a small number of key nodes in their systems, and deregulation allowed this growth to continue in a more streamlined and efficient manner. The air carriers would logically have had a small number of major maintenance facilities at the airports they served, and aircraft rotation schedules would be planned around these nodes. The problem for the airlines with the regulations enforced by the CAB was that the ability to add and cut routes was very limited, and so traffic flows and aircraft movements could not be channeled through such a narrow subset of nodes. The ADA did allow this streamlining to take place, and the hub-and-spoke networks currently being observed, are the logical progression for the airlines.

For all FAA hubs, it must be remembered that huge growth was taking place in the volume of traffic at these airports, in order for them to maintain their status (i.e. for medium hubs, they consistently needed to carry between 0.25 and 0.99% of total U.S. traffic). As with the large hubs, the increase in the number of carriers serving these airports is quite dramatic after 1977. For the medium hubs, the increase is at a slower rate, with this trend being consistent for the 10 years since 1977. The Gini index for the medium hubs is consistently lower than for the large hubs. The index began increasing substantially after 1977, rising from an average of 0.361 in that year to 0.567 in 1987. The airlines select these medium hubs as obvious centers of growth and accordingly, have used them as secondary hubs in their systems. Figure 3.6 plots the average Gini index for each hub type for the 18 year period, while Figure 3.7 shows the average number of carriers serving each hub type over the same period.

The number of carriers serving small hubs has increased also in the deregulated period. In 1969, there were an average of 2.4 carriers at the five small hubs in this sample. There were 2.6 on average in 1977 and 5.6 in 1987, indicating the improving role of the 'spoke' airports in the deregulated period; small airports are used by an increasing number of carriers as feeder routes into the main hubs. Looking at the traffic distributions at these airports, it is noted that only a small change has taken place in the

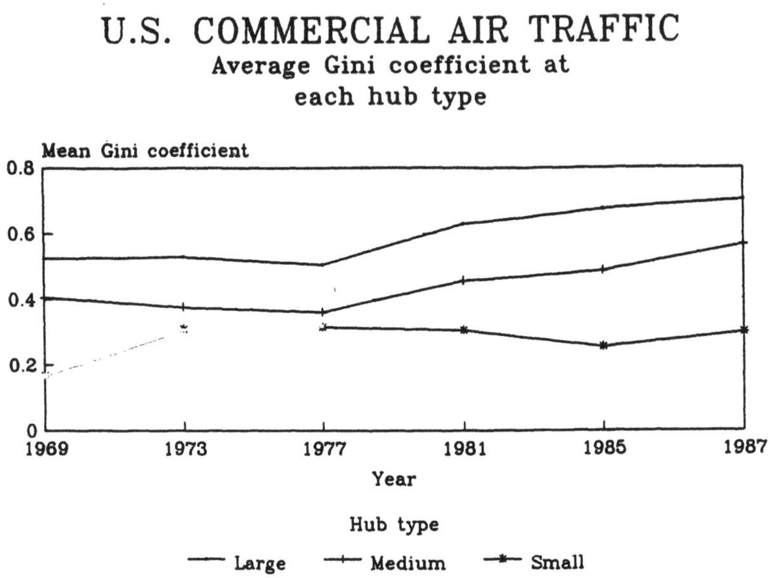

Figure 3.6

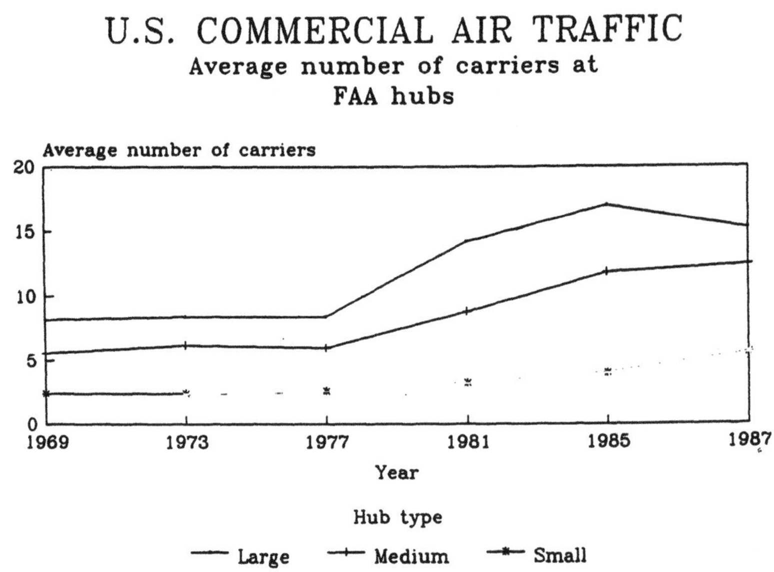

Figure 3.7

concentration levels, in the 10 years since deregulation. Generally speaking, the Gini index continues to reflect a more equitable distribution of traffic among carriers serving small hubs. This is good for these communities, since if their traffic load is being relatively evenly spread over an increasing number of carriers, they are also spreading their risks, giving them greater stability.

3.4 U.S. Air Carriers 1969-1988

Attention is now turned to the airline industry, and the size of that industry over the 19 year period, 1969-88. The Department of Transportation (DOT), (and before 1984, the CAB) grant a certificate of public convenience and necessity to certain air carriers to conduct scheduled air services, though charter and non-scheduled services may also be conducted. These carriers report operating and traffic statistics to the FAA twice a year, and summary details are presented annually in the FAA/CAB Airport Activity Statistics of the Certificated Route Air Carriers.

Figure 3.8 (page 57) plots the total number of certificated carriers for 1969-88. Between 1969 and 1978, the number of carriers fluctuated between 33 (1974) and 37 (1970), dramatically rising to 52 in 1979, the first fully deregulated year. The industry continued to expand in the subsequent years. In 1983, the total count dropped by 9 carriers, and the general trend recently has been a gradual fall-off in the number of carriers in the industry. The decline does not necessarily reflect financial failure of carriers, but rather the extent of mergers, takeovers and acquisitions by other airlines in the industry. The airline industry tends to be a high-capital, low profit business and the recent patterns in carrier numbers represent a rationalization in the industry in the more competitive, post-deregulation environment.

The FAA now classify air carriers on the basis of their annual operating revenues: since 1981, four carrier groups are reported, these are given in Table 3.7. The number of carriers in these groups are given in Table 3.8. Prior to 1981, 9 classes of air carriers were reported. These groups were

(1) Big four trunks (American, Eastern, TWA and United)
(2) Other trunks (trunks being pre-1938 entrants, serving primarily larger communities)
(3) Local service carriers (post-1938 entrants, operating routes of lesser density between the smaller traffic centers and the principal centers)
(4) Helicopter service
(5) Intra-Alaskan carriers (operating entirely within the state of Alaska)
(6) Intra-Hawaii carriers (operating entirely within Hawaii)
(7) All-cargo carriers
(8) Regional carriers
(9) Other carriers (carriers not fitting into any other category)

Table 3.7
FAA air carrier classification scheme since 1981

Carrier Group	Annual operating revenues
Majors	≥ $1 billion
Nationals	$100 million - $1 billion
Large regionals	$ 10 million - $100 million
Medium regionals	≤ $ 10 million

In order to make some kind of comparison, the first two categories were merged, and are compatible with the 'major carriers' of the 1980s. The local service carriers have been left as a separate group, while intra-Alaskan, Intra-Hawaii and the regional carriers have been combined as 'territorial carriers'. The 'other carriers' group in the classification scheme, was a catch-all category, and comparing 1980 with 1981 and 1982 data, it seems to refer to some national, but mostly large and medium regional carriers. The all-cargo and helicopter carriers are reported as separate classes. Table 3.8 tabulates carrier group trends.

The number of majors has remained relatively constant over the period 1969-88. The increase in the late 1980s reflects the inclusion of two all-cargo carriers (Federal Express and Flying Tigers). The greatest changes have occurred in the medium and large regionals categories. The number of medium regionals fell

Table 3.8
Number of FAA certificated carriers, by carrier groups, 1969-88

Year	Total carriers	Trunk[1]	Local service	Territorial & regional	Other	All-cargo	Helicopter
1969	34	11	9	6	1	3	4
1970	37	11	9	6	4	3	4
1971	36	11	9	6	3	3	4
1972	35	11	9	6	3	3	3
1973	35	12	8	7	2	3	3
1974	33	11	8	6	2	3	3
1975	34	11	9	6	2	3	3
1976	35	11	8	9	2	3	2
1977	35	11	8	9	2	3	2
1978	34	11	8	9	2	3	1
1979	52	11	9	8	19	3	1
1980	58	10	8	12	25	3	0

Year	Total carriers	Major	National	Large regional	Medium regional
1981	63	12	14	17	20
1982	64	12	15	11	26
1983	55	11	14	19	11
1984	62	11	19	23	9
1985	59	12	15	22	10
1986	55	13	20	18	4
1987	46	12	15	15	4
1988	41	12	11	15	3

Source: CAB/FAA data
1/ For explanation of carrier classes, see text.

sharply from 26 in 1982 to just 4 in 1987, while the number of large regionals has varied up and down from a high of 23 in 1984, to 15 in 1988. Looking at the trends for the earlier period (1969-80), there were consistent numbers in all categories up until 1979, when the 'other air carriers' category jumped from 2 airlines in 1978 to 19 in 1979, and 25 in 1980. The FAA classification scheme, reflecting their rigid control of the industry, broke down rapidly when regulations were removed in 1978. The major carriers served 77.4% of the total U.S. passenger traffic in 1987, and attention is now turned to examining two of these carriers in more detail.

Both Continental and American existed before regulation was first imposed in 1938, and as was pointed out in earlier chapters, airlines in this position were given special attention and protection under CAB regulation. The economic status of air carriers in the current period was, to a large extent, predetermined by their status under regulation. The trunk carriers emerged in the late 1970s with by far the largest share of the air passenger traffic market, the most extensive aircraft fleet and gate ownership levels and with long standing public reputations. In the ten years since 1978 however, three of the 11 trunks of 1978 have declared bankruptcy and two were acquired by other trunks. In 1989/90, 9 of the the 11 trunks were flying, many under different management and corporate frameworks. Table 3.9 summarizes the main changes taking place between 1978 and 1990.

3.5 Continental and American Airlines' Traffic Patterns, 1969-88

In this section, two major U.S. airlines are focused, the aims being to (1) examine the effects of deregulation on the size and share of two carriers' operations (2) trace the characteristics of these carriers' networks for a roughly 20 year period, and examine the changes in these carriers' networks in terms of their size, features of the airports served, traffic and distribution of traffic, and levels of concentration in the overall network system.

Table 3.10 shows a traffic index for Continental and American from 1969 to 1988 (1977, the base year, =100) and also the share

Table 3.9
Major changes in the U.S. trunk carriers 1977-1990.

1977 Trunks	Changes since 1977
American Airlines	1/
Braniff Airways Inc.	Bankruptcy declared 1982. Re-entered as a national carrier 1984 Operations ceased in 1990.
Continental Air Lines Inc.	Acquired by Texas International 1981. Bankruptcy declared 1983. 3/
Delta Air Lines Inc.	
Eastern Air Lines Inc.	Acquired by Texas International 1986 Bankruptcy declared 1989. Operations ceased in 1991.
National Airlines Inc.	Acquired by Pan American 1979.
Northwest Airlines Inc.	2/
Pan American World Airways	
Trans World Airlines Inc.	
United Air Lines Inc.	
Western Air Lines Inc.	Acquired by Delta Air Lines 1986.

1/ American Airlines acquired Air California 1987
2/ Northwest acquired Republic in 1986 (Republic consisted of North Central, Southern and Hughes Southwest)
3/ Texas Air (formerly Texas International) now consists of Continental Airlines (of which People Express (a post 1978 entrant), New York Air, Frontier, Rocky Mountain Airlines, Bar Harbor Airways, Britt Airways and Provincetown-Boston Airlines Inc are subsidiaries) and Eastern Airlines.

Table 3.10
Continental and American Airlines' traffic indices and share of domestic U.S. traffic market 1969-88

Year	Continental Airlines		American Airlines	
	traffic index	share of U.S. market	traffic index	share of U.S. market
1969	63.68	3.009	86.99	12.642
1973	90.03	3.443	92.48	10.874
1977	100.00	3.293	100.00	10.125
1981	111.49	3.048	107.92	9.071
1985	208.50	4.153	181.98	11.145
1987	534.82	9.099	242.29	12.675
1988	481.93	8.023	279.38	14.302

Source: CAB/FAA <u>Airport Activity Statistics of the Certificated route air carriers</u>, various editions.

of total domestic traffic handled by these airlines over the same period. American's volume of traffic has increased 280% since 1977, while its share of the passenger market is just slightly higher than it was in 1969. This airline's market share dropped in the 1970s and continued to decline initially under deregulation. American has managed to recapture its earlier share of the market via extension of its network and purchase of smaller feeder airlines (e.g. Air California Inc., and American Eagle). Continental Airlines was a smaller trunk, and its market share varied from 3.0% of total traffic in 1969 to 3.4% of traffic in 1973. The volume of traffic handled by Continental rose slowly after 1978. As seen in Table 3.9, Continental was acquired by Texas Air (formerly Texas International) in 1981, and declared bankruptcy in 1983. After re-organization in 1984 and 1985, the company purchased People Express (and its subsidiary, Frontier), New York Air and Rocky Mountain Airways Inc., from its parent company. Continental rapidly increased its volume of traffic, 530% over the 1977 volume. Its share of the U.S. market grew from 3.04% in 1981 to 9.1% in 1987, declining to 8% in 1988.

In Tables 3.11 and 3.12, the number of domestic airports served by both carriers are recorded for the period 1969-88, and these are subdivided into the FAA hub classifications. Also given in these tables, are the proportions of traffic at each hub type over the period. Both airlines have greatly expanded the size of their domestic networks since 1977, Continental increased its numbers from 33 airports in 1977 to 124 in 1988 while American increased its numbers from 50 in 1977 to 121 in 1988. Looking at the emphasis within the networks, American Airlines has consistently focused their operations on the FAA large hubs. The proportion of traffic at these large hubs increased in the deregulated period, when all FAA large hubs were served by this airline. The proportion of traffic at medium hubs has declined even though the airline served a larger proportion of FAA medium hubs in 1985, 1987 and 1988, then it did in 1977. These trends are illustrated in Figure 3.9.

Continental Airlines has significantly increased its traffic proportion at the large hubs it serves. In 1977, 67.8% of its traffic passed through large hubs, while in 1987, 79.6% of its

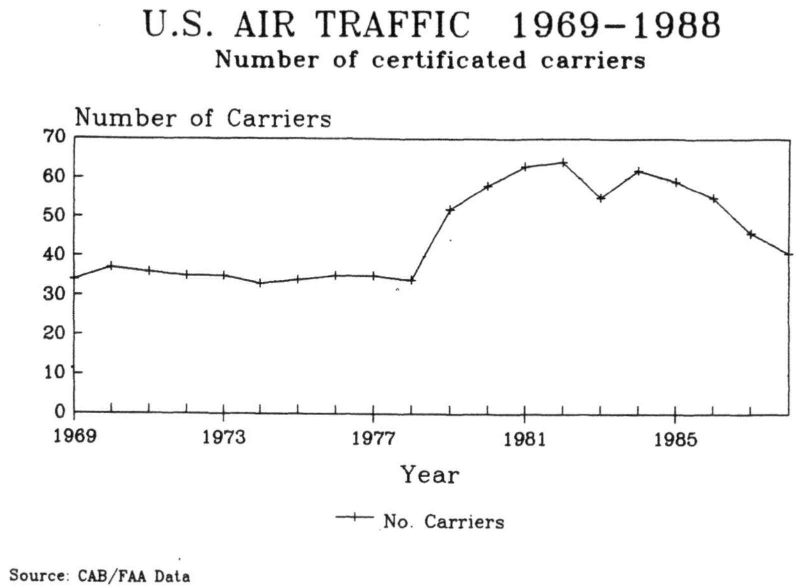

Figure 3.8

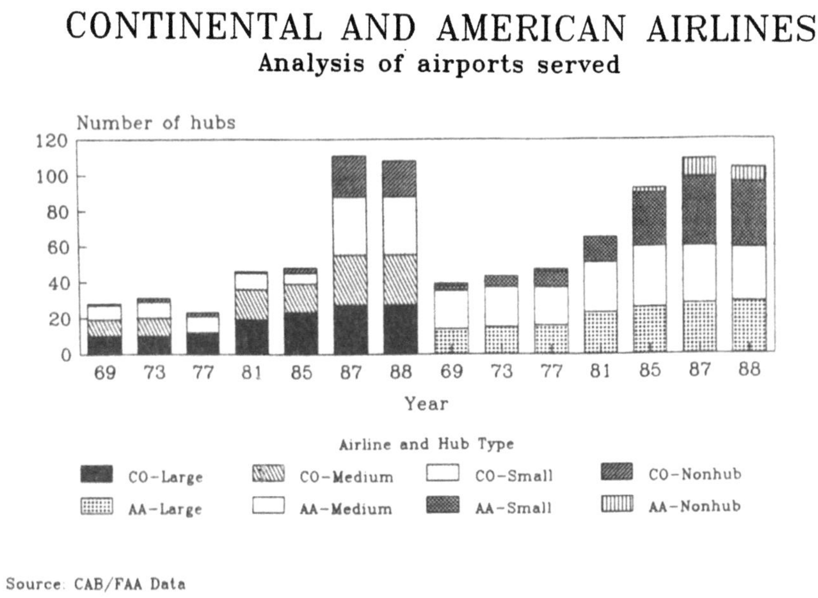

Figure 3.9

Table 3.11
Continental Airlines: traffic distributions and domestic system characteristics, 1969-1988

Year	Number of airports served	Number of hubs served				Proportion of traffic at each hub type			
		L	M	S	N	L	M	S	N
1969	28	10	9	8	1	63.35	23.37	13.09	0.19
1973	32	10	10	9	2	65.01	22.11	12.68	0.20
1977	33	12	9	9	2	67.81	20.49	11.92	0.07
1981	48	19	17	9	1	71.73	22.06	5.64	0.57
1985	54	23	16	6	3	83.37	11.84	3.89	0.89
1987	122	27	28	33	23	78.81	14.16	4.89	2.13
1988	124	27	28	33	20	79.60	14.39	4.27	1.74

Table 3.12
American Airlines: traffic distributions and domestic system characteristics, 1969-1988

Year	Number of airports served	Number of hubs served				Proportion of traffic at each hub type			
		L	M	S	N	L	M	S	N
1969	44	14	21	3	1	72.27	25.31	0.76	0.09
1973	47	15	22	6	0	71.47	25.46	1.43	0.00
1977	50	16	21	9	1	74.17	23.02	2.73	0.08
1981	67	23	28	14	0	79.85	17.94	2.22	0.00
1985	102	26	34	30	3	78.44	15.93	5.41	0.23
1987	119	28	32	39	10	72.69	20.98	6.01	0.32
1988	121	29	30	37	8	70.52	22.88	6.17	0.44

traffic was channelled through these communities, when all but one of the FAA large hubs were served. As with American the proportion of traffic at medium hubs has declined, along with a decline in the proportion of traffic at small hubs. In 1987 however, this airline dramatically increased the number of small hubs and nonhubs that it served: much of this increase in the network size is explained by the mergers and acquisitions mentioned above.

The traffic distributions over the airlines' systems were examined for each of the time periods selected. Scheduled and nonscheduled domestic passenger enplanements for each airport in the system were recorded for the six years. A series of summary statistics were computed for these traffic distributions for both carriers, and these are detailed in Tables 3.13 and 3.14. In Chapter 2, various network variables used in the economics literature were mentioned and these are now compared with the Gini index, Theil's entropy measure and the coefficient of variation. The two most popular series of measures used in the economics literature were the Herfindahl index and the proportion of traffic at the top 1,2 and 3% most utilized airports (hereafter called the McShan measures). In Tables 3.13 and 3.14, these measures are recorded for both airlines.

The Gini index, Theil measure and coefficient of variation all have one obvious major advantage over the McShan measure, since all of these measures utilize the full distribution, and therefore take into account the increases in the number of airports served by both airlines. Looking at one example, the fact that 16.4% of Continental Airline's total traffic was at its busiest airport in 1969, and 18.1% of its traffic was at its busiest airport in 1987, masks the 436% increase in the size of the system, as measured by the number of nodes. 18.1% of an airline's traffic at one airport, when it serves a total of 122 nodes, is intuitively more concentrated than having 16.4% of its traffic at one airport when only 28 nodes are served. In addition, any changes in the traffic distributions at the lower ranked airports will not affect the summary statistics computed by McShan and other researchers. The various measures in Tables 3.13 and 3.14 are plotted in Figures 3.10 and 3.11 to show the consistency of the three measures proposed here, in comparison with the rather

Table 3.13
Measures of concentration for Continental Airlines traffic distributions 1969-88

Year	C.O.V.	Theil index	Gini index	Traffic at top 1,2,3 & 4 airports				Herfindahl index
1969	1.185	-0.831	0.565	0.164	0.328	0.411	0.491	0.086
1973	1.220	-0.840	0.553	0.164	0.318	0.401	0.464	0.078
1977	1.255	-0.838	0.555	0.173	0.314	0.413	0.468	0.078
1981	2.252	-0.737	0.674	0.313	0.408	0.501	0.560	0.127
1985	2.382	-0.729	0.659	0.241	0.478	0.531	0.561	0.124
1987	2.763	-0.730	0.741	0.181	0.305	0.422	0.472	0.071
1988	2.764	-0.733	0.759	0.174	0.324	0.457	0.512	0.059

Table 3.14
Measures of concentration for American Airlines traffic distributions 1969-88

Year	C.O.V.	Theil index	Gini index	Traffic at top 1,2,3 & 4 airport				Herfindahl index
1969	1.276	-0.844	0.589	0.139	0.244	0.350	0.417	0.061
1973	1.387	-0.829	0.603	0.142	0.253	0.353	0.423	0.062
1977	1.525	-0.816	0.618	0.160	0.278	0.372	0.446	0.066
1981	2.157	-0.748	0.705	0.198	0.356	0.430	0.504	0.084
1985	3.255	-0.704	0.713	0.278	0.448	0.492	0.523	0.114
1987	3.205	-0.731	0.718	0.235	0.377	0.424	0.459	0.085
1988	2.720	-0.761	0.713	0.220	0.355	0.421	0.466	0.058

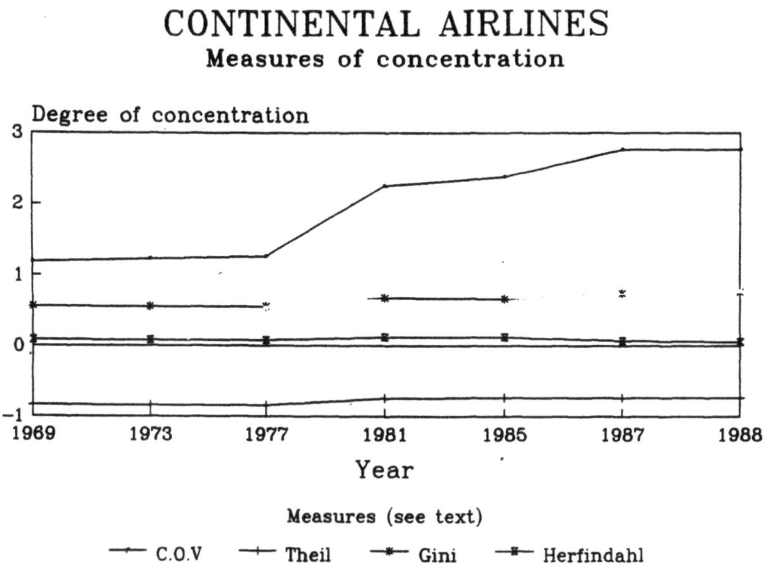

Figure 3.10

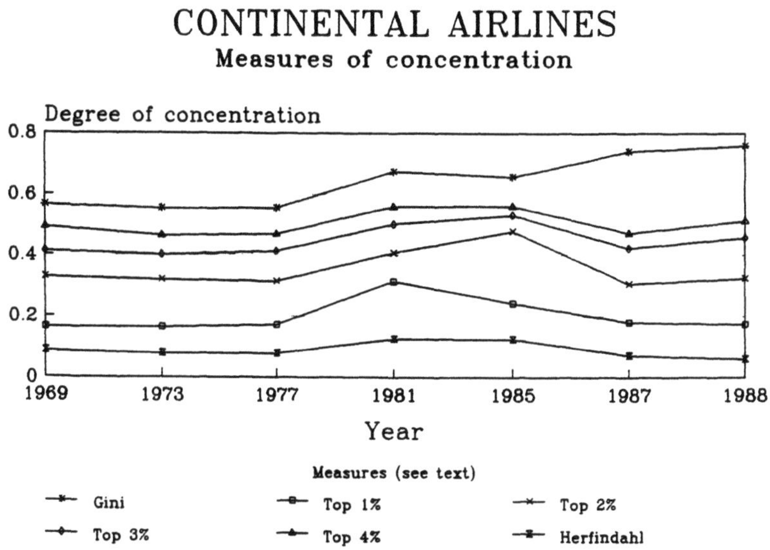

Figure 3.11

erratic patterns observed for the other measures.

Looking now at the more usual statistics computed to describe a distribution, the Gini index and Theil index are most commonly seen in the literature on income distribution. The coefficient of variation, Theil's entropy measure and the Gini index show gradual increases in both American and Continental Airlines' traffic distributions between 1969 and 1988. The three measures are generally consistent in the trends they show. In 1988, the three measures record further concentration in Continental's system, while in American's system for the same year the coefficient of variation and the Gini index record a slight reduction in concentration. The Theil measure on the other hand suggest a slight increase. The Herfindahl index shows a different trend however, the change in this index is much smaller over the period. It suggests very slight increases in concentration of traffic for both airlines from 1969-85, and a drop in concentration between 1985 and 1988 (even though, in the space of two years, there were 126% and 17% increases in the number of airports served by Continental and American Airlines respectively). The proportions of traffic at the top 1,2,3 and 4 airports served by both airlines, show the same trend.

In the analysis of income distribution, the poverty line is a pre-defined level of income, and income inequality is measured against that income level. In looking at air traffic distributions, the 'base-line' traffic level varies depending on the measure used. The traffic distribution measures at one of the large hubs in Table 3.4 were tested for their sensitivity to small changes (0.03%) in traffic proportions handled by successively higher ranked air carriers (larger changes affected the rank orderings). The effects of these changes on the four statistics are mapped in Figures 3.12 and 3.13. The Gini index shows a straight line relationship between the rank of the air carrier and the effect of the change on the index value. The Theil measure and coefficient of variation show close to log-linear relationships, though they are also affected by the actual proportion handled by each carrier. The Herfindahl index shows no specific pattern of change. All indices show declines in index values for lower ranked air carriers and increases in the indices for higher ranked

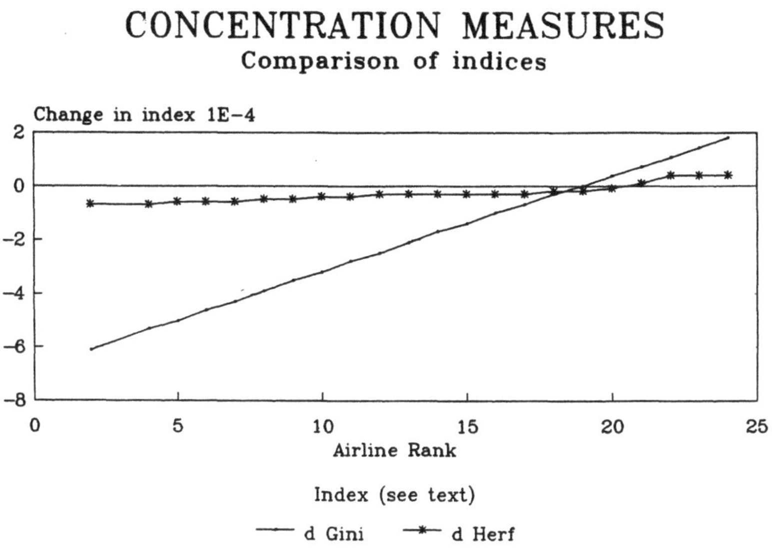

('d' refers to a change in the index)

Figure 3.12

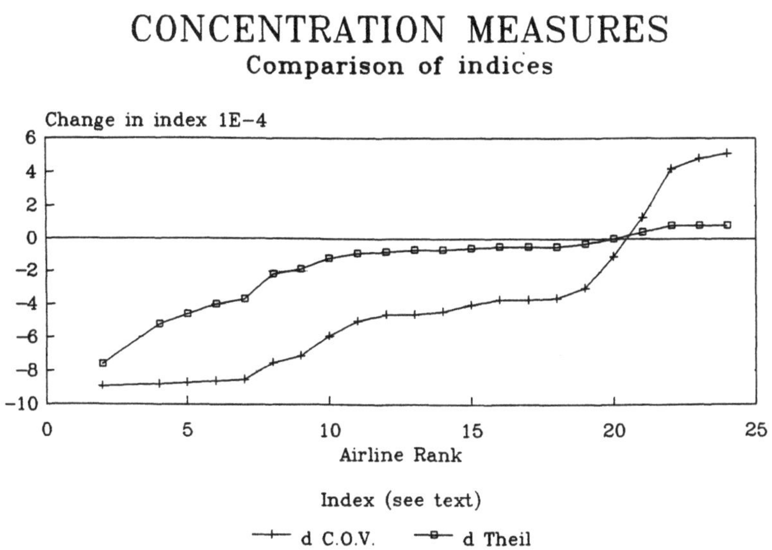

('d' refers to a change in the index)

Figure 3.13

values. The points in the distribution where the base-line traffic levels occur (i.e. rank where there is no change in the index), vary around the median of the distribution, with the Gini index base-line occurring just below the median, and all other measures occurring slightly above the median mark. The Theil and Gini indices are considered superior statistics for the reasons listed earlier and also because they make unit free comparisons. In addition, various decomposition schemes have been proposed for both of these measures: it is suggested here that further studies measuring carrier-specific network changes and their impacts on airport traffic patterns may aid the planning process at regional and national levels.

The Gini index, besides being a statistically and intuitively sound measure of concentration in a distribution, has the additional advantage of being easily illustrated. For this reason, it was selected in this book to illustrate concentration in airport and airline traffic distributions. Figure 3.14 illustrate the changing nature of Continental Airline's traffic distributions between 1969 and 1987. The high level of concentration in the airline's traffic patterns in 1969 and 1977 indicate that the recent phenomenon of hub-and-spoke networking is not so drastic a change in organization, but rather a refinement of a general trend among larger carriers to concentrate flows from an increasing number of nodes in their system, through a small number of key airports or carrier hubs. The same comments were made earlier in section 3.3, when airport patterns were examined. These trends have become exaggerated since 1977, but were very definitely in place prior to deregulation.

In conclusion then, the major U.S. carriers examined here have greatly expanded their networks (be it through internal growth or merger with/acquisition of other carriers) since 1977, and have maintained or increased their market shares by serving a very high proportion of the large and medium FAA hubs, and by streamlining their services through a small number of key hub-airports. The size of the national airways system serving the scheduled traffic market has contracted considerably since deregulation. This smaller national network which is handling a

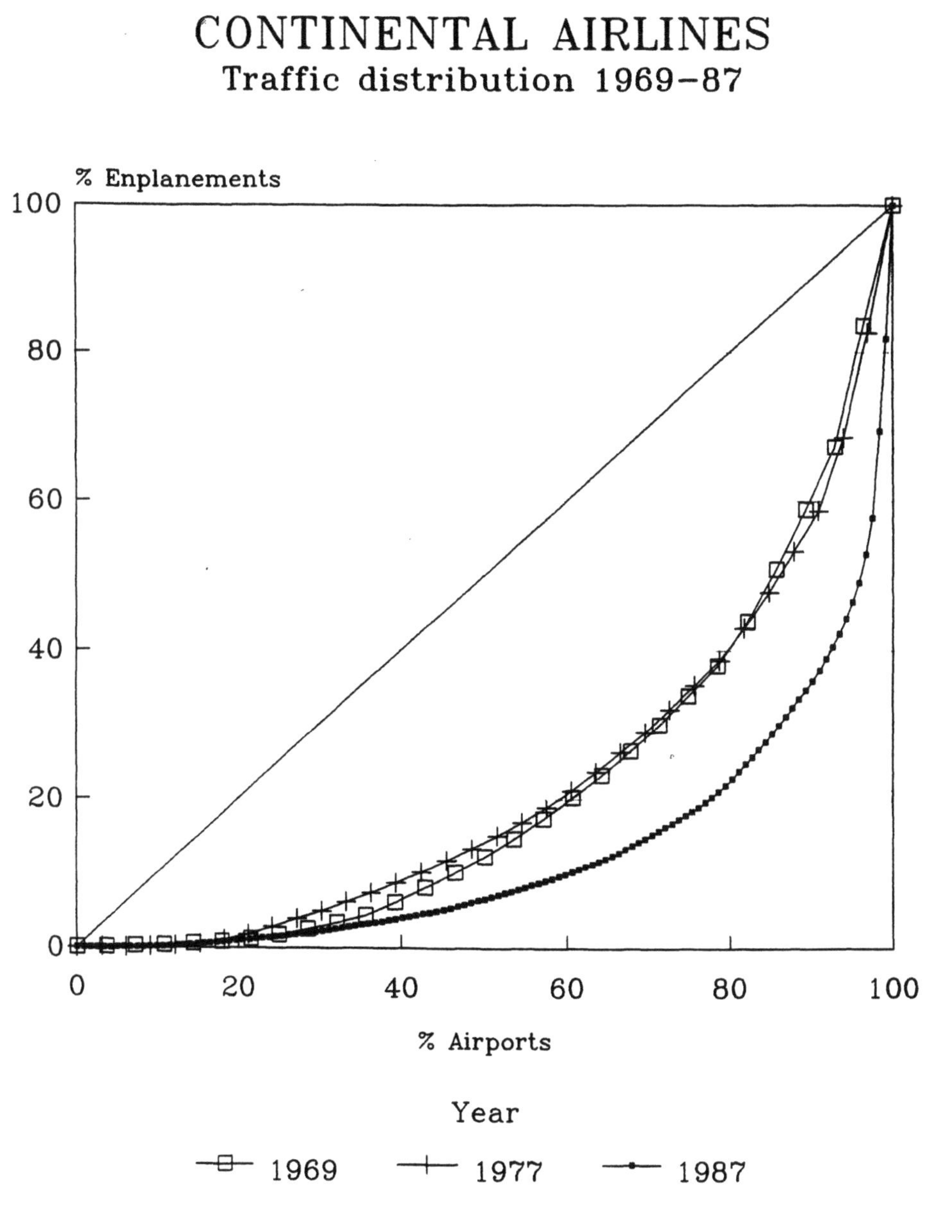
CONTINENTAL AIRLINES
Traffic distribution 1969-87
% Enplanements
100
80
60
40
20
0
0
20
40
60
80
100
% Airports
Year
1969
1977
1987

Figure 3.14

greatly increased volume of traffic, is now being better serviced by the major carriers. The 11 trunk carriers in 1977 accounted for c78% of all U.S. traffic in that year. In 1988 the 12 major carriers accounted for 77.40% of U.S. traffic. Change in the industry structure has occurred among the smaller carriers; the large and medium regionals were the classes of scheduled carriers experiencing the greatest change in terms of new entrants and subsequent exits.

This chapter proposed the Gini index as a superior summary measure of airport and airline traffic concentration, compared with the measures previously used in the economics literature. It was shown how concentration in the traffic patterns of both the major airlines and the large airports was at a high level prior to deregulation, and since 1978, the patterns have gradually become even more concentrated. Previous measures proposed to examine these kinds of network changes have failed to highlight this important historical trend, and researchers have heralded the 'hub-and-spoke' route systems as a purely post-deregulation phenomenon. The same measures tracing traffic patterns back over a 20 year period show very slight variability in concentration levels; it was shown how insensitive these measures were to changes in the size and nature of the traffic distribution. It is recommended that the Gini index or Theil's index be used in subsequent studies of national or industry network analysis. Discrepencies in research results in this area are less likely if the nature of the network is better understood, and, if accurate and consistent measures are utilized.

In the next chapter, the scale of analysis is reduced to focus on the operation of a single airline firm, and to demonstrate how this important network component can be included in an economic model of the airline firm.

CHAPTER 4
DEVELOPMENT OF A PRODUCTION MODEL OF THE AIRLINE FIRM

In this chapter a model of the airline firm is developed which takes account of the joint-product nature of airline production over a system of routes. In Chapter 2, the airline's schedule construction problem was briefly outlined and the view of the network as both the production plan and the product was explained. Using this view of the network, it will be shown how characteristics of the routes may be included in the firm's production function in a logical and formal manner. This analysis is detailed in section 4.1. The multi-product production model is then developed in section 4.2. In section 4.3, an alternative view of the airline production problem is put forward. This involves an examination of the problem at the airport level rather than at the route level. The disadvantages and difficulties of this approach are highlighted and it is recommended that the route approach be used instead, despite the fact that data at the airport level is more easily available. The empirical production model is then explained in section 4.4 and the dependence between routes in the network is catered for by use of a spatial autoregressive structure in the estimating procedure. In section 4.5, the 'frontier' or 'best practice' production model is set out and the estimation procedure explained.

4.1 The Airline Network - Defining a Network Variable

In Chapter 2, the detailed and complicated structure of the airline's schedule construction problem was outlined. It was shown that for large networks (of 50-100 nodes, and 200 or more routes), these problems were impossible to solve optimally, so that practitioners in this area relied on the development of heuristics to solve smaller sections of the problem.

In order to put forward an economic model for evaluating the airline's network schedule, it is necessary to simplify the decision making process and assume that the firm takes the following steps. The first decision involves selecting airports or nodes to serve. These points in the plane form the basis on which the network is planned. For the airline industry in the United States up until 1978, the selection of airports was determined for the most part by two key factors: (1) the fact that all of the founders of airline companies based their operations in the nearest large regional population centers of the continent and then developed around these points. After regulation was imposed on the industry in 1938, as discussed in the previous chapters, the CAA and CAB heavily influenced the entry patterns for new markets and thus heavily influenced the airports and routes that each airline served. (2) After 1978, the airlines were given considerable freedom to add or cut routes from their network and thus to expand the size and scale of their domestic networks. The position of the airlines after 1978 was however greatly influenced by its position prior to 1978, and the large trunk airlines had substantial advantages at the large airports. The number of airports served by the 'major' U.S. carriers increased rapidly either by internal growth or by merger or acquisition.

The airline's decision to offer service from a new airport depends on such factors as the traffic potential for the airline at that center, the competitive position of the airline, the airport facilities and availability of gate slots, and the relative geographical location of the airport and its suitability as an additional node in the current network. It is assumed here that this set of nodes is given, and that the second and third steps are the main focus for the economic analyst.

Having selected a set of airports to serve, the next major decision that needs to be made is the determination of which routes (or arcs between the nodes) to serve. From the economist's point of view this is the key step to be taken before production levels may be determined. A simple model of route selection will be set out here. The doubly-constrained Gravity model is generated as a solution to this route selection procedure.

Route selection will be determined by using an extended version of the 'Transportation Problem'. The famous transportation problem of linear programming involves finding the minimum cost (c_{ij}) or minimum distance (d_{ij}) involved in moving passengers/freight from origins to destinations subject to additivity constraints or conservation of flow constraints. The transportation problem in the context of airline scheduling would involve finding the minimum distance routes for satisfying passenger flows from origin airports to destination airports. This problem yields (2n-1) routes for n airports and thus gives the minimum possible number of routes or the most efficient set of routes for the objective in hand. By including an entropy constraint which requires a greater degree of accessibility than the minimum distance solution would yield, a gravity model may be derived [Wilson (1970), Erlander (1978), Hansen (1974)]. This entropy constraint allows for a greater number of routes in the optimal solution (the maximum possible number being $[(n^2-n)/2]$), with each possible route in the system representing a plausible production possibility for the airline firm.

The general form of the gravity model is

$$T_{ij} = A_i O_i B_j D_j \exp(-\eta\, d_{ij})$$

$$\text{with } A_i = [\sum_j B_j D_j \exp(-\eta\, d_{ij})]^{-1}$$

$$\text{and } B_j = [\sum_i A_i O_i \exp(-\eta\, d_{ij})]^{-1}.$$

where T_{ij} is the expected number of journeys/passengers moving from i to j, and

$$O_i = \sum_j T_{ij}$$

$$D_j = \sum_i T_{ij}$$

This is the doubly constrained gravity model with an exponential distance (or cost) function. The model may be extended to include a weight variable, W_{ij}, with parameter Θ_1 and a simplified distance function (with parameter Θ_2) to give

$$T_{ij} = A_i O_i B_j D_j W_{ij}^{\Theta 1}\, d_{ij}^{\Theta 2}$$

$$\text{with } A_i = [\sum_j B_j D_j W_{ij}^{\Theta 1}\, d_{ij}^{\Theta 2}]^{-1}$$

$$\text{and } B_j = [\sum_i A_i O_i W_{ij}^{\Theta 1}\, d_{ij}^{\Theta 2}]^{-1}.$$

These traffic potentials (T_{ij}s) or 'route potentials' are inputs in the firm's next decision process, which involves maximizing the output that can be produced on the routes for a particular budget. The firm will select the optimal input combinations which are technically feasible, and which maximize it's profit. These input-to-output relationships on each of the routes are captured in the firm's multi-product production function. By expanding the firm's production problem to include a network variable measuring each route's traffic potential, an airline production function may be estimated and used to assess route efficiency in terms of its technical performance. This network variable captures specific characteristics and features of each route and may be thought of as a product differentiating variable.

4.2 The Airline's Production Function

The airline's production function specifies a relationship between the units of air transport services produced on each route (Q) and the inputs labor (ground and flight crews), vehicles, fuel and also a network variable representing the way or path input. In a spatial setting each possible route in the system represents a plausible production possibility for the airline firm. The airline selects a very small subset of these routes based principally on the traffic potential that the route has to offer. The network variable may be thought of as an asset to the firm, measuring the route potential for generating passenger traffic. As each route's attractiveness (i.e. potential traffic to be generated from both nodes) increases, the level of output on the route is expected to increase also.

In the time series model of production, a time variable is specified by merely increasing the time variable for each successive period (usually one year or one quarter) by one unit since the time increment between measurement points is exactly the same. Movement is in one direction (either increasing or decreasing) so that what happens in time 't' can affect what will happen in time 't+1' but not vice versa. In a spatial setting production is taking place over a two-dimensional plane. In a fixed time period movement can take place in any direction and activity at the point (x,y) can affect or be affected by activity at points close to, or even at substantial distances from it. The location of the node in a network is fixed so that the physical relationship between any pair of points will not change. There need not be a regular interval between the points; typically in fact, there is an irregular spacing. The production of flight services over space depends on the traffic potential existing between pairs of points and the intervening distance between them. By including a variable in the production function which captures this traffic potential, the nature of the network may be included in the model.

In addition to this direct passenger potential which characterizes each route, attention must also be given to the route's indirect passenger potential. Indirect passenger potential refers to the expected rate at which passengers flying on one route connect to or from this route to others. This connection rate reflects the joint-product nature of production over the system of routes and it is both of these potentials which help the airline to select the routes to serve, and also help to organize production on the selected routes. This organization may be viewed as technology over space leading to improved economies of scope and in order to capture this factor, indicators of interaction potentials need to be defined along with methods for integrating the production process over the system of routes. By capturing this 'dependence' among routes in the system, the economies of scope available to the firm in joint-production will be included in the model. However it is not possible to measure the extent of economies of scope obtained.

The production function represents the maximum output obtainable from a given set of inputs at fixed factor prices and is the solution to the firm's profit maximization problem for fixed costs. In the airline's case, the production functions to be estimated take the following general form:

Y = f(PILOT, FA, FUEL, CAP, AGENT, LOAD, NETWORK)

where Y is a vector of outputs, measured by passenger seats available per route. PILOT and FA are pilot and flight attendant crew flying time per route. FUEL is the total quantity of the fuel input on each route. CAP is the capital input on each route (measuring the aircraft's relative capacity and operating cost ratio). AGENT is the number of ground agent time units allocated to each route. LOAD is the average load factor for each route or airport, (capturing seat capacity versus seat utilization ratios on each route [see Caves et al. (1984)]). NETWORK is a variable measuring the passenger potential on each route and this variable is generated by calibrating a doubly-constrained gravity model with weight and distance variables.

4.3 An Alternative View of the Airline's Production Process

An alternative approach to estimating a production function for a single airline firm would involve treating each airport as a separate production center and estimating the production function by using data from each airport as production activity associated with that center. The production function would be the same as above except that the variables are now redefined as follows: OUTPUT represents the vector of flight services produced from each airport in the network; PILOT and FA are the pilot and flight attendant crews associated with production from each airport (half of the pilot flying time associated with each route would then be assigned to the originating airport and the other half assigned to the destination airport); LOAD is the average load factor associated with flights from the airport; AGENT and CAPITAL are the ground crew hours and vehicles inputs associated with the airport and NETWORK represents characteristics of the airport which are attributes to the firm in producing flight services. The network variable focuses on spatial characteristics of the airport's relative location. There are difficulties in incorporating additional information about traffic potential for the airline at each airport. Reynolds (1989) tested two network variables in such an airport model. These were generated by using 'second order analysis' [see Getis and Franklin (1987)] and 'trend surface analysis [see Johnson (1978)]. Second order analysis produces an expected level of interaction within a specified distance based on population data and on the relative location of each airport (measured by planar co-ordinates). Trend surface analysis also produces an expected interaction level at each airport based on the position of the airport in the network. The results of this analysis at the airport level highlighted the problems of aggregation in production analysis. The nature of 'hub-and-spoke' networking means that a large proportion of an airline's traffic occurs at a very small number of airports (usually three or four 'hub' airports). In a regression analysis, these airports completely dominate the estimating process and accordingly dominate the resulting parameter estimates. Figure 4.1 illustrates the dominance of the Denver, Houston and Newark hubs in Continental Airlines' route system.

Intuitively the route analysis provides a richer explanation of the production process for the airline. For example, if two carriers both serve the same six airports, the carriers may have completely different networks based on the way that the airports are connected. So in describing the two networks we are more interested in the way the points are connected (i.e. the routes served) rather than which points are included. Furthermore, air transport involves the production of flight services on a route: to assign that production to one of the two nodes does away with the nature of the transport industry's product. It is therefore recommended that at the firm level, analysis should focus on production on each route as opposed to production from each airport.

4.4 Functional Form of the Airline Production Function

In selecting a functional form for the production function, a translog model was initially chosen. The translog model is a flexible functional form, meaning that it satisfies the mathematical definition of a local second order approximation [see Berndt and Christensen (1973)]. These functional forms do not restrict the form of distribution, substitution or scale effects of a technology in the way that first order forms do. The first order forms carry with them a series of maintained (and non-tested) hypotheses. The flexible forms on the other hand have a smaller number of maintained hypotheses and have become more popular in applied economic analysis in recent years. Because of serious statistical problems and difficulties with the data set, a simpler functional form had to be used in the production analysis and in the analysis of efficiency [see Reynolds (1989) for extensive discussion of results and problems with the translog specification].

By imposing restrictions on the nature of the firm's

CONTINENTAL AIRLINES

Traffic distribution 1987/88

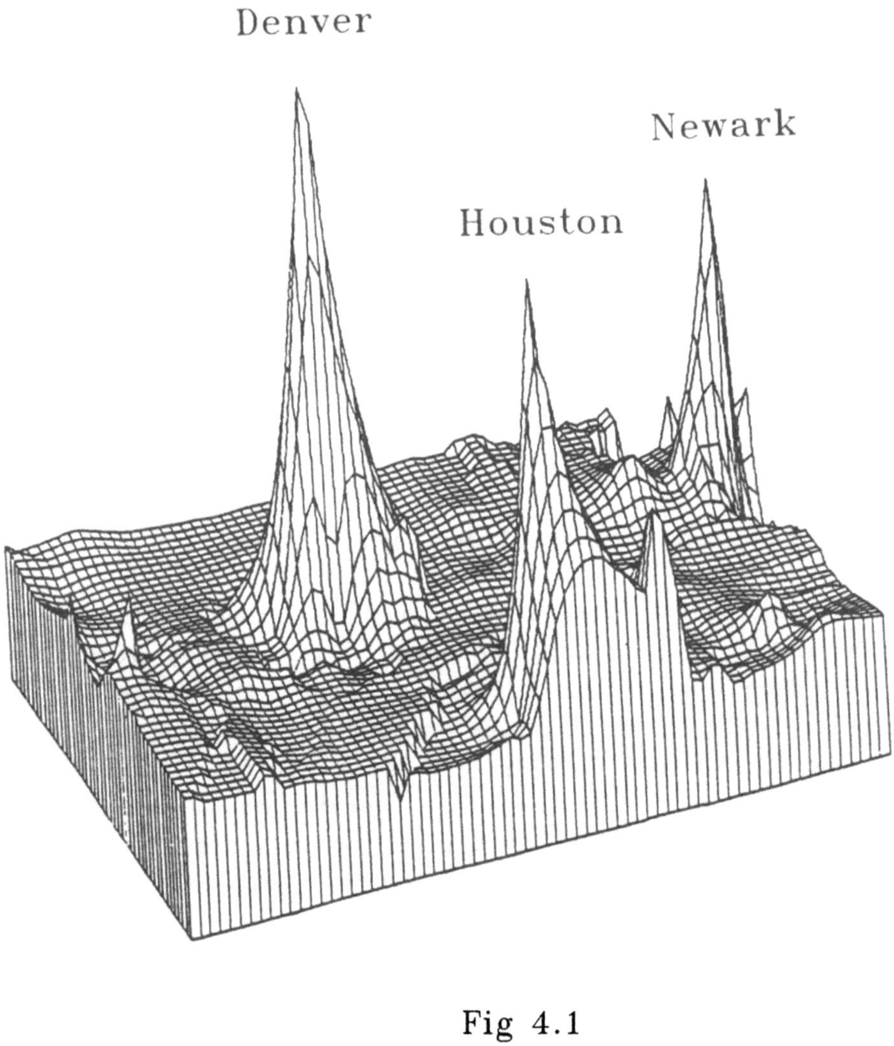

Fig 4.1

production process, the translog function collapses to a log-linear Cobb-Douglas production function. The Cobb-Douglas production function is a first order, log-linear model relating output to input levels. Using Zellner, Kmenta and Dreze's formulation (1966), the expected output level (E(Q)) is given as

$$E(Q) = A\, L^{\alpha_1} K^{\alpha_2} e^{1/2\sigma^2}$$

so that the parameters of L and K (two inputs) are independent of the error term. We note that the constant A is dependent on the error term when the variables are converted to log form [Goldberger (1968)].

For the airline firm, the Cobb-Douglas production function for the routes will be of the following form:

$$ln\ Y = \alpha_0 + \alpha_1 ln\ \text{PILOT} + \alpha_2 ln\ \text{FA} + \alpha_3 ln\ \text{FUEL} + \alpha_4 ln\ \text{CAP}$$
$$+ \alpha_5 ln\ \text{AGENT} + \alpha_6 ln\ \text{LOAD} + \alpha_7 ln\ \text{NETWORK} + u$$

where u, the error term, is initially assumed to be normally distributed, and the other variables are as described earlier. In this model, returns to scale are calculated as $\Sigma\alpha_i$. These are aggregated returns to scale which are a weighted average of route-specific returns to scale, magnified by the degree of economies of scope in production. As Willig (1979) points out, it is not possible to directly measure the economies of scope to the firm in this kind of analysis, since this would require production data for single-route airlines, each producing on just one route in the network. No such firms exist and accordingly measurement of economies of scope gained from production by one firm over a network is not possible.

Economies of scope in production exist for the airline if the firm can produce more output over a series of routes for a given budget than many firms each producing on just one route for the same total cost outlay. So let

$$\mathbb{Q} = [Q_1, Q_2] = \mathbb{Q}(L,K)$$

where $\mathbb{Q}$ represents an output vector for two products jointly produced by one firm, using input factors L and K. Let the following functions represent the production functions associated with production of the two goods separately by two firms:

$$Q_1 = Q_1(L,K) \qquad \text{and} \qquad Q_2 = Q_2(L,K).$$

Economies of scope in production are defined as S_c where

$$S_c = [Q_1(L,K) + Q_2(L,K) - Q(L,K)] / Q(L,K).$$

Let S_1 and S_2 be the product specific returns to scale for the two goods Q_1 and Q_2. Then the overall returns to scale, S, are given as

$$S = \frac{w_1 S_1 + w_2 S_2}{(1-S_c)}$$

where w_1 and w_2 are weights based on the proportions of Q_1 and Q_2 produced.

Serious statistical problems encountered when using a number of computed variables (computed from the data set provided by Continental Airlines) required that a composite "equipment" variable be formed from the PILOT, FA, FUEL and CAP variables. Thus further analysis used four input variables rather than the seven specified above. The first four inputs are combined into an equipment variable (EQUIP) using Principal Components Analysis (PCA), so that the production function is of the form

$$ln\ Y = \alpha_0 + \alpha_1 ln\ \text{EQUIP} + \alpha_2 ln\ \text{AGENT} + \alpha_3 ln\ \text{LOAD}$$
$$+ \alpha_4 ln\ \text{NETWORK} + u.$$

A review of the method of Principal Components Analysis is given in Chapter 5, when the generation of the EQUIP variable is discussed. One of the advantages in using PCA is that coefficients for the four original variables may be derived from the estimated coefficient by using the factor loadings. It is expected that all of the α_i coefficients will have positive signs, though in the case of the load factor, the sign may be positive or negative since it represents a quality of service variable from the passengers point of view (i.e. higher load factors imply lower quality).

To estimate this production function as it is given above implicitly assumes that production of flight services on each route is independent of production on other routes. This is not the case however since it is assumed that economies of scope exist for the firm implying that production on one route <u>does</u> influence production on other routes. To explain the form of this dependence between routes in the network, we remember that the passenger potential on each route needed to be included in two ways. First a variable should be included that would capture the expected level of traffic on each route for the airline, given competition on the route, the populations of the two centers served and, the relative location of the centers. This effect is captured by the gravity model. Second, the passenger transfer rates from one route to another should be taken into account. While a route may not have a high traffic potential relative to many other routes in the system, the opportunity for that route to provide transfer passengers to other routes is an important factor which the airline needs to consider. This transfer rate makes it possible to use larger aircraft on routes with high levels of local demand (i.e. high direct demand from the airport's hinterland) and substantial numbers of passengers connecting from other routes. This is a major source of economies of scope for the airline, since there are route-specific scale economies associated with the use of larger aircraft. In terms of estimating a production function with this connection potential included, it was decided to use a spatial autocorrelation weight matrix (which records passenger transfer levels) to try and account for the additional variation in production or output levels not catered for by the route's input levels. The spatial autoregressive model has a covariance matrix of the form

$$\sigma^2 \, [(I-\rho W)'(I-\rho W)]^{-1}$$

where W is a matrix representing the spatial dependence between all pairs of routes, and ρ is the autoregressive parameter [see Anselin, (1989)]. With OLS estimation of the parameters in the Cobb-Douglas function (α),

$$\alpha = (X'\Omega^{-1}X)^{-1} \, X'\Omega^{-1}Y$$

and since in OLS estimation Ω is assumed to be the identity matrix,

$$\alpha = (X'X)^{-1} \, X'Y$$

where Y is the output vector, X is the matrix of inputs, α is a vector of parameter estimates $\alpha_1, \ldots, \alpha_n$. By including the connection rates in the estimation process, the parameters to be estimated are of the form given above,

$$\alpha = (X'\Omega^{-1}X)^{-1} \, X'\Omega^{-1}Y,$$

with the variance-covariance matrix given by

$$\text{var}\,(\alpha) = \sigma^2 \, (X'\Omega^{-1}X)^{-1}.$$

Since Ω^{-1} is not an identity matrix, (Ω^{-1}= $[(I-\rho W)'(I-\rho W)]$), OLS will produce biased estimates of the parameters. Generalized Least Squares (GLS) is the appropriate estimating method, and it will be used in estimating the production functions when connection rates are included. The connection rates are on a scale of zero to unity and these will be included directly as the omega matrix. It is implicitly proposed, in using the connection rates in this way, that they represent the spatial relationships between sets of routes and the relative importance or strength of these relationships (as measured by the scaling parameter ρ) in the production of flight services. The GLS procedure will produce an estimate of this scaling parameter ρ, which, combined with the connection rates will allow for the computation of the W matrix, representing the structure of spatial dependence in the airline's route system. This procedure is equivalent to the weighted least squares method.

When estimating the Cobb-Douglas functions, the omega matrix will be included in a GLS estimation procedure. Ideally the weight matrix should consist of information on route by route connection rates. Data provided by Continental Airlines included only the average connection rates to or from each route for all other routes in the system, rather than specific route by route rates. Accordingly, the omega matrix consists of a diagonal matrix with average connection rates on the main diagonal and zeros elsewhere.

4.5 Frontier Production Functions and the Measurement of Technical Inefficiency

The estimation of a frontier production function amounts to specifying a one-sided error term in the estimating model, and then using either nonlinear or maximum likelihood methods to estimate the boundary functions. The classification of frontier technology depends on the way the frontier is specified and estimated. The two methods that will be used in this study are both classed as deterministic statistical frontiers, estimated by corrected ordinary least squares (COLS), and by maximum likelihood with an assumed Gamma distribution on the error term [see Richmond (1976), Førsund, Lovell and Schmidt (1979)].

The COLS method assumes that the error term, u, has a finite mean μ and variance σ^2: for a Cobb-Douglas function then,

$ln\ y = \alpha_0 + \sum_i \alpha_i\ ln\ X_i - u$ and

$ln\ y = (\alpha_0-\mu) + \sum_i \alpha_i\ ln\ X_i - (u-\mu)$,

so that the new error term has zero mean. This error term satisfies all the usual assumptions except normality. Best linear unbiased estimates (BLUEs) of $(\alpha_0-\mu)$ and the α_is can be obtained by OLS: the $E(u_i)$ can be estimated and used to adjust the OLS estimate of the intercept. COLS provides consistent estimates of all of the parameters of the frontier function.

The second method utilizes maximum likelihood to estimate the Cobb-Douglas function for the spatially uncorrelated variables, with a gamma distribution assumed for the error component. Aigner, Lovell and Schmidt (1977) have pointed out that the maximum likelihood (ML) estimates are completely dependent on the form of the disturbance. The maximum likelihood estimates will have their standard statistical properties under two conditions for z (where $u=e^{-z}$):

(1) the density of z is zero at z=0

(2) the derivative of the density of z with respect to its parameters approaches zero as z approaches infinity.

These conditions are not satisfied under all assumed distributions. The gamma distribution does exhibit both of these conditions for n>0. This does not, however, constitute a firm *a priori* reason for selecting this distribution, but it was selected for purely statistical reasons nevertheless. Estimation using ML in this manner is equivalent to the quadratic programming solution for the convex hull of the output distribution.

The frontier models were estimated for the routes data that include the influence of spatial interdependence measured via the connection rates. It was decided to specify the deterministic frontier function where errors were assumed to have a gamma distribution principally because the major aim in this research in estimating a frontier function is to allow for distinction between those routes which are more efficient and those which are lessefficient. Other methods such as Schmidt's two-component error structure only give an average measure of inefficiency for the system of routes [Schmidt (1976)].

The COLS method was applied to the spatially uncorrelated data set (estimated using the weights from the GLS procedure). The COLS equation is thus

$$ln\ (Y-\rho Y) = ln\ A + \sum_{i} \beta_i\ ln\ (X_i-\rho X_i) + u,$$

where $u=e^{-z}$ and the probability density function of z has a gamma distribution [Richmond (1976), Førsund et al. (1979)].

The second method used a maximum likelihood function to estimate the Cobb-Douglas function for the spatially uncorrelated variables with a Gamma distribution assumed for the error component (MLEγ). These production functions were estimated for the system of routes and for each of four subsets of routes using the two methods just described. The two methods produced very similar results in terms of the measurement of efficiency of each route in the system. Rather than compare frontier output and observed output in their given measurement units, it was decided that measuring efficiency in a standardized efficiency unit would be a more useful exercise, and that comparison of results would be easier across subsets of routes. In addition, the frontier function depends to a large extent on the position of the positive outliers. The most efficient market in the system was an infrequently used route, and comparison of all other routes with this outlier would be misleading. Accordingly, the frontier output level was compared with the observed output level, and the mean difference (i.e. the mean level of inefficiency) was computed along with the standard deviation of the difference. For each route, the level of inefficiency was then converted into a standard normal deviate. In the tables and discussions in Chapter 6, efficiency is thus measured in standard deviation units around the mean level of inefficiency.

In this chapter, a model of the airline's production function was developed. The inclusion of a new variable capturing the route characteristics was justified and integrated with the production model in a logical way. It was shown that this variable represents the optimal solution to the firm's network design problem and once the decision has been taken as to which routes to offer services on, the firm can then decide the levels of output to produce on each route. It was proposed that the indirect passenger potential, generated on other routes by passenger transfers, should be included in the production model by using a GLS estimator with weights made up of connecting passenger levels.

As was explained in Chapter 2, because of the complexity involved in formulating and trying to solve the airline scheduling problem, there is a high possibility of technical inefficiency in the system. Frontier production analysis can aid schedule planners in evaluating the schedule by highlighting the characteristics of the inefficient routes and identifying the factors which improve performance of specific routes and of the entire system. Once the production model has been appropriately specified and adjusted to take account of the spatial dependence between routes, this measurement of efficiency can be successfully achieved. In Chapter 6 the results of this procedure applied to Continental Airlines' domestic route system are presented and discussed. In the next chapter, details of the data used in the analysis are given.

CHAPTER 5

DATA DESCRIPTION AND COMPUTATION OF THE PRODUCTION VARIABLES

In order to test the production model with a network variable and measure the degree of inefficiency in the system as described in Chapter 4, detailed route by route production data was required from a large U.S. airline. Continental Airlines were approached, and a request was made to them for route statistics for a one year period. The airline obliged, and provided an extensive data set for their domestic system operations on non-stop markets for the year 1987/88. The airline staff pointed out that they did not maintain extensive historical records on their routes: the information was kept for two years at most. In addition, in view of the dramatic changes in this particular airline's size and extent, much of the historical information is now irrelevant. The airline's schedule planning framework operates for periods much shorter than 12 months, so the staff suggested that there would be inherent biases in the one year data set. This data set is described in detail in section 5.1. In section 5.2 the generation of input and output variables for the production analysis is described. Since four of the input factor variables were highly correlated, it was decided to use Principal Components Analysis to generate a new single variable representing most of the common variation in the four factors. This procedure is described in section 5.3. The final section of the chapter deals with the specification and calibration of a doubly-constrained Gravity model, which is used as the 'network' variable in the production function.

5.1 Data Provided by Continental Airlines

Continental Airlines provided a listing of their domestic fleet for the month of February, 1988, along with each aircraft's rate of fuel consumption per flying hour (recorded per minute in Table 5.1), and the crew size, differentiated by pilot crew and flight attendant crews. This information is summarized in Table

Table 5.1
Continental Airlines aircraft characteristics 1987/88

Aircraft	Fuel per minute	Pilot crew	Flight attendants	Capacity (seats)	Capital ratio
DC9-10	12.50	2	3	83	0.269199
DC9-30	13.33	2	3	108	0.296900
DC10-10	39.17	3	9	284	0.736299
DC10-30	45.83	3	9	217	0.688099
A300B	30.00	3	8	272	0.582499
MD80	14.58	2	4	143	0.346100
B737-100	11.92	2	3	94	0.297199
B737-200	12.50	2	3	91	0.327699
B737-300	11.92	2	4	130	0.410899
B727-100	19.50	3	4	119	0.375099
B727-200	20.67	3	4	150	0.409500
B747-200	62.00	3	14	398	1.000000

5.1. For each route or market segment served, and for each aircraft type on these routes, total passengers, total departures, and total number of available seats were given over the one year period from July 1, 1987 to June 30, 1988. The load factor for each aircraft type, on each route was also included. There was a total of 1705 entries in this data set, data for the 338 domestic U.S., Canadian and Mexican routes served by the airline in the 1987/1988 period. The airline operated out of 128 airports. The Canadian and Mexican routes were dropped from the data set, as the analysis in this book focuses purely on U.S. patterns and networks. There are 296 routes for the U.S. in the time period under consideration, and these are operated from 113 airports.

A great circle mileage chart for the domestic route system was given by the airline. For each airport served, the total staff employed was listed for each of 14 categories. The agent/clerical staff was itemized in this table.

Continental furnished three full sets of the company flight timetables for June 1987, January 1988 and May 1988. These gave the arrival and departure times for each route, by equipment (i.e. aircraft) type. Using these timetables, the time block per aircraft type was averaged in both directions (there was, on occasion, up to a 50 minute differential in the west-east flight times compared with the east-west flight times), for the Summer and Winter seasons, and the average flight time in minutes, by equipment, was recorded for each route. These times were used in computing the production variables, since they include a measure of delay on each of the routes: the FAA require that 75% of an airline's flights arrive within 15 minutes of the published schedule times, so it is argued that these block times capture the additional outlay in producing flight services on routes with greater delays at one or both nodes of the segment. Finally, Continental provided a listing of the average proportion of passengers connecting to and from each route, for an average month in the year (February, 1988) and for the Summer months on the 'Summer Only' segments.

Continental Airlines domestic route system over the period 1987/88 consisted of 296 routes between the 113 airports that they served. In Chapter 4, the FAA hub classification scheme was explained, and this scheme is now used to classify Continental's airports and routes served. Table 5.2 tabulates the number of each hub type served and the percentage of Continental's total traffic at these hub types. The table then shows the route classification scheme, consisting of large to large routes (i.e. routes where both the origin and destination airports were large FAA hubs), large to medium routes (routes where one node was a large FAA hub and the other, a medium hub), large-small routes, large-nonhub routes, medium-medium routes, medium-small routes, medium-nonhub routes, small-small routes, small-nonhub routes and nonhub-nonhub routes. The number of routes in each category is tabulated, along with the proportion of traffic, the average length of routes, the average load factor, average connection rates, and, average distance-time ratios.

In Chapter 3, it was shown how total U.S. traffic was heavily concentrated at the FAA large hubs. The airlines examined, concentrated their operations at these airports. Looking at the route by route traffic data, it is not surprising that almost one half of all of Continental's traffic, measured by the number of departures, takes place between pairs of large hubs. An additional 33% of the airline's traffic comes from large to medium hub markets. A further 9.9% is based on flights between large and small hub markets, and 4.6% between large and nonhub markets. Thus just under 97% of the airline's total traffic in the 1987/88 period, passed through at least one large FAA hub airport.

Looking at the average distance for each of the market types, Table 5.2 shows a generally decreasing mileage as the markets get thinner. The overall average distance per route for the airline was 729 miles for the 1987/88 period.

Table 5.2
Characteristics of Continental Airlines' airports and routes, using the FAA hub classification scheme.

FAA hub type	Number served	Proportion of traffic
Large	35[1/]	72.885%
Medium	29	18.140%
Small	31	5.761%
Nonhub	17	3.214%

Market type	No. Depts	% Depts	No. routes	Average distance	Load factor	Connect rate	D.T.R
LL	254629	48.98	104	945.98	.5424	.4426	5.793
LM	172526	33.19	106	714.65	.5045	.4674	5.395
LS	51293	9.87	37	532.94	.4846	.6209	5.148
LN	23863	4.59	18	452.90	.4464	.6712	4.791
MM	7318	1.41	12	436.19	.4022	.7066	4.642
MS	1063	0.20	3	423.17	.5688	.8589	4.415
MN	0	0.00	0	-	-	-	-
SS	2010	0.39	5	139.90	.4848	.9670	3.105
SN	4351	0.84	6	143.34	.3385	.9093	3.603
NN	2832	0.54	5	132.11	.2688	.9772	3.774
Total	519885	100.00	296	729.03	.5046	.5214	5.367

1/ These figures refer to the number of airports in each FAA category: New York for example, is an FAA large hub, and Continental Airlines serve 2 airports in that hub community. The figure above includes both of these airports.

The load factor gradually decreases as the routes get thinner. The average load factor for Continental's system is 50.5%, it being higher for large to large hub routes, at 54%. The connection rates (meaning the proportion of passengers transferring to or from another route) get progressively higher as the markets become thinner. Obviously most of the passengers on these routes will connect to heavier markets. On the very thin routes for instance, the nonhub to nonhub, or nonhub to small routes, where flight distances are very short, the airline provides a series of short hops to the bigger centers. Load factors are low, but these services act as 'passenger gathering' trips, feeding into the more lucrative markets. These routes, the 'spokes' in the network, are a critical part of the system's successful operation, and it is clear from the load factor and connection characteristics that the firm faces a different set of objectives and tradeoffs for these routes.

The distance-time ratios (D.T.R.) for the various market types, measures the average number of miles covered per minute on the different route types. This ratio declines generally as the routes become thinner. This is due to two main factors: first of all, the fact that the heavier routes make use of faster, larger aircraft, since these routes are generally covering longer distances. Secondly, gate times at both nodes become a more significant component when the flight distances are shorter.

It was decided to examine the performance and utilization of different aircraft types on Continental's system of domestic routes. Three different classifications of equipment were used. First, statistics for each type of aircraft were analyzed by market type. Second, the number of different equipment types on each route was counted, and used to distinguish different groups of markets. Finally, aircraft were distinguished by the distance range that they are commonly classified into (e.g. using Janes All The World's Aircraft); there are three groups, long haul, medium haul and short haul aircraft. In addition to these classifications of Continental's equipment, the routes were divided into five groups, based on the airline's designated hubs. The groups are (1) routes connected to or from Denver, Colorado (Continental's major domestic hub) (2) routes connected to or from Houston (IAH), Texas (Continental's second hub) (3) routes connected to or from Newark, New Jersey (Continental's newest and now third hub) (4) the three routes between Denver, Houston and Newark (5) other routes in the system, not connected directly to the three main hubs. Using the full data set (i.e. the 1621 entries covering each of Continental's routes and equipment types), a series of tables were extracted for various combinations of equipment classifications and route or market type categories.

Continental's domestic fleet consists of mostly medium haul

aircraft (61.7%). The short haul aircraft account for 33% of the fleet and long haul aircraft account for the remaining 5.3%. Most of the long haul aircraft, (wide-bodied jets, such as the B747-200, DC10-10 and DC10-30), are used very infrequently on the domestic routes. These aircraft are presumably used on the heavier routes for the small proportion of non-stop transcontinental flights, and also between the airline's hubs, as part of the equipment rotation process for the international market. Continental's international traffic for 1987 accounted for 7.1% of their total traffic in that year. This international traffic is channeled relatively evenly through many of the FAA's large hubs, particularly in the southern parts of the U.S., since Continental's international system focuses heavily on the Caribbean and Latin America. Houston (IAH) handled 17% of the airline's international traffic in 1987, while Newark handled just over 5%. It must be pointed out however that the airline had an agreement with the European airline SAS (Scandinavian Air Systems) for European routes: Newark's low international traffic level may mask considerable transfer rates to European routes served by this other carrier. The long haul airliners handled 3.89% of Continental's domestic departures in the 1987/88 period; 94.5% of departures were on routes where a large pool of equipment types were available (i.e. at least 7 aircraft types). The routes utilizing these aircraft were predominantly long distance routes. Table 5.3 shows the distribution of departures over the various sets of route and equipment categories. The long haul airliners are used mostly on routes into/out of Newark, making them available for Newark's international traffic market.

In all distance classes, routes with greater numbers of equipment types available have consistently better load factors than routes with a smaller number of equipment types, and indeed, load factors increase as market distances increase. This would indicate a better matching of equipment to market segments of longer distances, and with heavier passenger levels. The distance-time ratio is improved over successively greater distances, presumably for the reasons given earlier, and also, it is speculated, because of this improved matching of equipment types on routes, given their passenger potential and mileage.

The load factors are higher in all distance categories using short haul aircraft. These aircraft are efficient over most of the domestic routes operated by Continental, given that 54.6% of routes are of 500 miles or less, 37.8% of routes of are between 750 and 1500 miles, and the remaining 7.6% of routes are of between 1500 and 2565 miles. The airline's scheduling problem then comes down to matching passenger demand or passenger potential at different times of the day (and indeed on different days) with the aircraft capacity of the equipment available on the route. Table 5.4 tabulates average load factors over various sets of routes. The connection rates for the system of routes show a different trend. Table 5.5 details connection rates for 4 sets of cross tabulations. The connection rates are higher on the shorter distance routes, and also on routes with the smallest number of different aircraft types available. Presumably, on short routes (which are generally thinner markets) one or two short range aircraft regularly service the markets. The connection rates are lower also for long-range equipment types, but as was pointed out, these aircraft service mostly long distance routes, and where they are used on shorter length routes, this is presumably done to relocate large aircraft.

Routes were divided up on the basis of the major hub that they were directly connected with. Average connection rates are highest for those operated out of Denver. Routes out of Houston have relatively similar connection rates to Denver, Houston's Intercontinental airport (IAH) being Continental's second hub. Newark's connection rates are substantially lower overall, particularly on routes to other large or medium FAA hubs (23.2% and 26.8% respectively of total departures for that airport) compared with Denver and Houston. These rates suggest that Newark, as far as Continental's passengers are concerned, is seen more as an origin or final destination in the Northeast region, rather than a transfer point. The distinguishing point about an airline's hub is that the connection rates, and connection potential for flights into the airports are high; the low rates at Newark could be explained a number of different ways. First, the Northeastern regional sub-market is a relatively new focus of attention for Continental, their hub at Newark coming about with the purchase of

People Express in 1984. The low connection rates may indicate the very slow process, for Continental, of integrating or merging a new regional hub into an existing system. At the same time, Continental is in the process of reducing their traffic levels at Dulles airport in Washington D.C., and the influence of this airport's traffic levels on the Northeast may help explain Newark's unusually low connection levels. In addition, Continental is competing for traffic in this new sub-market, and possibly experimenting with new connection segments out of Newark. The routes out of Newark at the moment are mostly of distances of less than 500 miles. Another reason for the low connection rates might be the fact that the data reflects connections within the domestic system only; Newark and neighboring New York City handle very large levels of international traffic, and the low domestic connection rates may hide connections to other airline systems or international routes.

The load factors are significantly higher than average on Continental's three sets of hub routes, averaging roughly 52% for the hub routes, and 44% for other routes not directly connected to the Denver, Houston or Newark. It was noted already that load factors were lower on the thinner routes, but that the majority of passengers on these routes connect to other heavier routes. There are more equipment types available on these airline hub to hub routes.

Clearly, there are striking differences in the airline's performance, operation and characteristics at the different hubs, and for the 'spoke' routes. To summarize, it has been noted that Denver-based routes had lower load factors, higher connection rates, and the majority of these routes were of moderate stage length (500-1000 miles). Houston-based routes were evenly divided between short, moderate and long distance segments, and load factors were consistently above the company's average for all the distance-based classes of routes.

Newark's performance, in terms of the connection rates achieved on its related routes, did not seem to be typical of a large airline's third hub. The possible reasons for this were noted above. The average load factors achieved on these routes were the highest for the airline, and they were particularly high in the moderate- to long distance segments (i.e. greater than 750 miles). Traffic levels between Newark and the other hubs were at

Table 5.3
Percentage of total departures for crosstables of distance categories with groups of aircraft range, amounts of different equipments available, and Continental hub airport markets

Distance in miles	Long range	Medium range	Short range	Total
0 - 249	0.490	10.285	7.912	18.688
250 - 499	0.060	14.038	7.767	21.866
500 - 749	0.074	9.169	4.796	14.039
750 - 999	1.305	12.926	7.558	21.790
1000 -1499	0.769	11.262	3.951	15.982
1500 -2565	1.119	4.017	2.500	7.636
Total	3.817	61.698	34.485	100.000

Distance in miles	Number of different equipment types available			Total
	1 - 3	4 - 6	7 - 10	
0 - 249	0.347	4.941	13.400	18.688
250 - 499	0.348	3.595	17.922	21.866
500 - 749	0.297	4.245	9.497	14.039
750 - 999	1.065	8.782	11.942	21.790
1000 -1499	0.297	7.111	8.574	15.982
1500 -2565	2.214	3.916	1.506	7.636
Total	4.568	32.590	62.842	100.000

Distance in miles	Denver	Houston	Newark	Interhub	Other	Total
0 - 249	2.153	5.057	7.639	0.000	3.838	18.688
250 - 499	4.604	4.311	6.242	0.000	6.709	21.866
500 - 749	7.988	2.204	2.125	0.000	1.722	14.039
750 - 999	12.302	3.884	1.829	1.506	2.269	21.790
1000 -1499	3.828	6.460	2.733	0.807	2.153	15.982
1500 -2565	2.750	1.548	1.306	0.845	1.187	7.636
Total	33.626	23.463	21.873	3.158	17.878	100.000

Table 5.4
Average load factors for crosstables of distance categories with groups of aircraft range, amounts of different equipments available, and Continental hub airport markets

Distance in miles	Long range	Medium range	Short range	Total
0 - 249	0.474	0.417	0.466	0.438
250 - 499	0.342	0.436	0.461	0.442
500 - 749	0.541	0.503	0.530	0.515
750 - 999	0.571	0.544	0.553	0.548
1000 -1499	0.479	0.589	0.558	0.569
1500 -2565	0.514	0.533	0.550	0.532
Total	0.489	0.499	0.515	0.505

Distance in miles	Number of different equipment types available			Total
	1 - 3	4 - 6	7 - 10	
0 - 249	0.398	0.389	0.478	0.438
250 - 499	0.397	0.409	0.460	0.442
500 - 749	0.398	0.504	0.539	0.515
750 - 999	0.415	0.524	0.588	0.548
1000 -1499	0.499	0.566	0.580	0.569
1500 -2565	0.529	0.523	0.558	0.532
Total	0.455	0.590	0.524	0.505

Distance in miles	Denver	Houston	Newark	Interhub	Other	Total
0 - 249	0.445	0.499	0.468	0.000	0.405	0.438
250 - 499	0.475	0.473	0.464	0.000	0.395	0.442
500 - 749	0.505	0.576	0.521	0.000	0.471	0.515
750 - 999	0.550	0.544	0.652	0.602	0.487	0.548
1000 -1499	0.512	0.567	0.646	0.410	0.577	0.569
1500 -2565	0.623	0.553	0.472	0.546	0.471	0.532
Total	0.523	0.536	0.532	0.517	0.442	0.505

Table 5.5
Average connection rates for crosstables of distance categories with groups of aircraft range, amounts of different equipments available, and Continental hub airport markets

Distance in miles	Long range	Medium range	Short range	Total
0 - 249	0.531	0.728	0.739	0.723
250 - 499	0.497	0.488	0.501	0.494
500 - 749	0.444	0.516	0.524	0.516
750 - 999	0.529	0.533	0.537	0.534
1000 -1499	0.423	0.392	0.412	0.401
1500 -2565	0.283	0.348	0.373	0.339
Total	0.429	0.520	0.540	0.521

Distance in miles	Number of different equipment types available			Total
	1 - 3	4 - 6	7 - 10	
0 - 249	0.972	0.810	0.636	0.723
250 - 499	0.594	0.539	0.465	0.493
500 - 749	0.572	0.532	0.497	0.516
750 - 999	0.332	0.545	0.538	0.534
1000 -1499	0.282	0.412	0.407	0.401
1500 -2565	0.299	0.386	0.289	0.339
Total	0.476	0.551	0.503	0.521

Distance in miles	Denver	Houston	Newark	Interhub	Other	Total
0 - 249	0.669	0.613	0.407	0.000	0.897	0.723
250 - 499	0.644	0.587	0.283	0.000	0.483	0.494
500 - 749	0.566	0.579	0.300	0.000	0.475	0.516
750 - 999	0.624	0.568	0.343	0.540	0.355	0.534
1000 -1499	0.551	0.481	0.202	0.390	0.315	0.401
1500 -2565	0.433	0.491	0.184	0.470	0.275	0.339
Total	0.592	0.552	0.290	0.464	0.580	0.521

relatively low levels, and this fact combined with the previously noted low distance/time ratios seem to indicate a greater stress on service purely in the Northeast regional submarket.

It was decided, in view of the differing characteristics on four obvious subsets of routes, and the relatively large number of observations for each subset, that separate production functions for each category would be estimated. It was felt that, in this way, more light would be shed on the airline's production process, and that comparison of the resulting estimates would yield additional useful information. The four subsets of routes consist of (1) routes connected directly with Denver, (2) routes connected directly with Houston, (3) routes connected directly with Newark and (4) all other routes in Continental's domestic system.

5.2 Computed Input and Output Variables

In addition to the data described in the previous section, capital ratios were required for each aircraft type in order to generate the production variables. Continental Airlines were unable to provide this information, so a 1983/84 CAB publication, Aircraft operating costs and performance report (1984), was consulted (this publication was discontinued after 1985, and these costs statistics are no longer published in FAA or DOT annuals). This publication lists each aircraft's average operating cost per block hour, broken down by maintenance, depreciation, fuel, crew, leasing and other operating expenses. These operating cost ratios were used in conjunction with a capacity ratio for each aircraft type used by Continental, to compute a capital stock weighting. These were then used to estimate the capital inputs for each route.

The data set described in the previous section consisted of widely differing variable ranges and scales. It was decided to reduce the size of the variable ranges by normalizing departure rates on each of the routes, and producing average sums for each characteristic per departure. This also had the effect of solving a heteroscedasticity problem. To compute the agent labor inputs for each route, the total agent labor minutes for the year was computed for each airport. It was assumed that all agents worked

47, 40-hour weeks per year. These totals were divided by the total number of departures out of each airport, and allocated to a route based on its share of total departures. The agent input was then summed for the two airports comprising the route, averaged and multiplied by the normalized number of departures. This was used as the agent labor input (AGENT).

The number of passengers, and number of available seats per aircraft type, per route, were multiplied by the flight block time, to obtain output variables, PAXTIM and STSTIM respectively and then averaged and indexed by the number of departures on the route. The load factor per route was the average load factor per indexed departure for each route (LOAD).

Pilot flying time per departure per route (PILOT) was calculated by taking the flying time (block time) per route per equipment type and adding this time to a preparation time. The preparation time consisted of a basic 15 minutes plus an additional 15 minutes for each hour flying time. This total for route and equipment type was then multiplied by the number of pilots per aircraft, and the number of departures per aircraft. A weighted average of the number of pilot minutes for the route was computed, and this figure was finally multiplied by the normalized number of departures for that route.

The size of the flight attendant crew on each flight, depends on the number of passengers on that flight. It was decided then, that the flight attendant crew for each aircraft type should be split up into a minimum number and prorated number of attendants (i.e. the number of attendants based on the load factor). This figure was computed for each equipment type, for each route. The flight attendant minutes was computed by taking the block time and preparation time (used above, to compute pilot minutes), and multiplying this sum by the number of flight attendants and the number of departures, and then averaging this for the route. The figure was multiplied by the normalized number of departures (FA).

It was decided to prorate a small proportion of the aircraft's fuel consumption by the load factor for each equipment type and route. This fuel consumption rate was then averaged for the route, and multiplied by the normalized number of departures (FUEL).

The capital weightings were multiplied by the route distance, and then adjusted by the distance-time ratios for each equipment type for each route. The weighting per mile was averaged for all equipment types and recorded for each route (the 'CAP' variable). It was found that there was a high degree of correlation between this CAP variable and the FUEL, PILOT and FA variables. The next section describes how this problem was handled.

5.3 Principal Components Estimates of the 'EQUIP' Variable

The correlations between the four variables, PILOT, FA, FUEL and CAP were particularly high. The simple correlations between each of the four variables are recorded in Table 5.6. They indicate a serious multicollinearity problem.

Table 5.6
Simple correlations between PILOT, FA, FUEL, and CAP variables

	PILOT	FA	FUEL	CAP
PILOT	1.000	.966	.949	.896
FA		1.000	.993	.974
FUEL			1.000	.982
CAP				1.000

Solutions to multicollinearity problems include Principal Components Analysis, estimation via ridge regression, or using ratios or first differences. While the latter method more often introduces heteroscedasticity, the former method is flawed in that it cannot be justified theoretically, or indeed statistically. Principal Component Analysis (PCA) offers a much more reasonable approach to solving the problem, by incorporating much of the common variability in the data into a smaller set of components. The main drawback with this technique is that the components often have little economic interpretation. However, in this particular case, there is a strong justification for combining the four variables, measuring pilot, flight attendant, fuel and capital inputs, since these collectively capture the equipment assignment for the routes. It is argued that the biggest single set of decisions for the airline (given that the actual routes have been selected) is the equipment allocation process. By computing a single component from the four inputs, it is argued that an equipment variable is derived, and may be validly included in the model. For the four variables, in each of the route subsets, a single component was computed and used in the production function estimations.

The method of principal components is often suggested (though rarely used) as a possible solution to the problem of multicollinearity. PCA is a purely mathematical device producing a set of mutually orthogonal or uncorrelated "components" derived from the original set of variables. Maddala (1977) describes the method in the following way: If there is a set of k explanatory variables, then a series of linear functions of these variables,

$$I_1 = a_1x_1 + a_2x_2 +, \ldots, + a_kx_k$$

$$I_2 = b_1x_1 + b_2x_2 +, \ldots, + b_kx_k \text{ etc.}$$

may be computed, subject to constraints on the coefficients a_i and b_i. The a's are chosen such that the variance of I_1 is maximized subject to the condition that $\sum_i a_i^2 = 1$ (the "normalization condition"). In this case, I_1 is said to be the first principal component, and is the linear function of the x's with the highest variance. The linear function I_2 requires that the b's be chosen so that I_2 is uncorrelated with I_1 and has maximum variance subject to the condition $\sum_i b_i^2 = 1$. I_2 is said to be the second principal component. There are k such linear functions of the x's which will account for the total variation in the x's. The method is useful if a subset of the I's are used, since otherwise, the answers will be the same as those for the full set of x's.

The main problems with the method lie in the fact that

(a) the first principal component may not be the one most correlated with the independent variable, even though it picks up most of the variance of the x's.

(b) the principal components produced often have little economic meaning or interpretation. This question has been addressed above already.

Table 5.7 shows the high degree of variability in the four variables captured by the first component in the analysis. The non-hub routes have the lowest score in this regard with 96.9% of the variability expressed in the component. Clearly then, the high correlations are consistent across all sets of routes. The factor loadings show very little variation in terms of the contributions of each variable to the first component. For the three hub routes, the flight attendant variable has the highest loading, while for the non-hub routes, the capital variable has the highest loading. The variability in the factors is relatively greater (20%), principally because of the larger number of cases in these groupings.

Table 5.7
Factor loadings for variables used in the principal components analysis.

Set of routes	Factor loadings on variables FUEL	PILOT	F.A.	CAPITAL	Proportion of variability in first component
All routes	-28.21	-30.53	-32.29	-33.17	.9654
Denver	-10.11	-10.58	-11.32	-10.80	.9709
Houston	-12.01	-11.95	-12.83	-12.35	.9659
Newark	-14.24	-13.84	-15.09	-13.54	.9650
Nonhub	-16.32	-18.51	-18.98	-21.08	.9591

Given the relatively similar factor loadings, and the high degree of variability explained by the first component, it is argued that using this single component in the rest of the analysis is justified. It is also proposed that this component can be thought of as a logical 'equipment' variable, since the four original variables, PILOT, FA, FUEL and CAP, depend heavily on the equipment selection decision.

5.4 Gravity Model Estimates, and the Route Attraction Variable

Various forms of the gravity model were used to generate alternate measures of accessibility or attraction for the routes in Continental's system. Earlier, in Chapter 4, the general form of the gravity model was specified. The models investigated in this analysis, included origin-constrained models, a doubly-constrained model and several unconstrained models for each hub. The models set out a relationship between the number of interactions occurring between nodes i and j (Tij) and (1) the population of i and j weighted by the number of commercial airports in each city/community, and, the number of other certificated passenger carriers serving both airports, (2) the distance between i and j, and, (3) the local versus connecting traffic on the route.

A series of multiple regression equations were estimated to examine the strengths of the relationships between the independent variables and the levels of interactions (Tij). The variables used in this preliminary investigation, are outlined in Table 5.8.

For the total domestic system operated by Continental Airlines, there were very poor relationships between passenger traffic on each of the routes, and both the population of the two airport hinterlands, and the intervening distance. It was shown in Chapter 3, that traffic levels at large and medium hubs particularly, varied considerably among carriers. The summary

Table 5.8

Investigation of the relationships between independent variables and the levels of interaction in Continental Airline's domestic system, 1987/88.

Variable	Description
STS_{ij}	Total number of Continental seats on each route
POP_i	1980 population of community i
POP_j	1980 population of community j
NC_i	Number of other carriers serving community i
NC_j	Number of other carriers serving community j
$DIST_{ij}$	Great circle distance between i and j in miles
$CO\%_i$	Continental's proportion of total traffic at i
$CO\%_j$	Continental's proportion of total traffic at j
$CONNECT_{ij}$	Connecting passenger ratio to route i,j
$STSLOC_{ij}$	Total number of locally originating passenger seats on each route. (STS_{ij}-(STS_{ij} x $CONNECT_{ij}$))

Simple adjusted R^2 between STS, STSLOC, DIST, and various population weights

Weight	$DIST_{ij}$	STS_{ij}	$STSLOC_{ij}$
$\sqrt{POP_i.POP_j)/(NC_i.NC_j)}$	.0334	.0654	.0240
$(POP_j)/(NC_j)$	.0004	.0250	.0162
$\sqrt{POP_i.POP_j)/(CO\%_i.CO\%_j)}$	.0915	.2136	.1682

measures of traffic distributions showed considerable variation in the proportion of traffic handled by carriers serving these communities. Accordingly, while it may be expected that total interactions on all of the routes in the U.S. show some degree of association with local population and intervening distance, this relationship is not apparent for just one firm's traffic compliment. Factors such as the airline's 'hub' airport locations (and therefore, rates of connection among passengers), concentration in regional markets and mergers or agreements with smaller airlines will all act to distort the degree of association between the variables. Table 5.6 demonstrates this breakdown in the relationship between Continental's interaction levels and the populations of each origin-destination pair, weighted by the number of additional certificated passenger air carriers serving the nodes (it is implicit from the weighting scheme, that on average, each carrier is assumed to serve an equal share of the market).

The locally originating traffic on each of Continental's routes was also examined using simple multiple regression estimates. Again, it may be seen from Table 5.8, that the relationships are very weak. The next step was to look at Continental's share of traffic in each market, and use this proportion to prorate the population statistics. This procedure takes into account Continental's hub locations, and relatively greater emphasis on the mid-continental submarkets. Regressing this new weighted population variable with distance on Continental's traffic levels, was more successful. The R^2 value was .36 for the entire system of routes. It was decided that this particular population variable would be most suitable for the gravity model calibration, since it took account of Continental's relative emphasis in the total U.S. air traffic market.

The intervening distances and average weighted population for each route were then used to calibrate the gravity model. The gravity model may be formulated in a number of different ways, depending on the aim of the researcher. For example, the problem may be set out so as to predict as accurately as possible, the level of interaction on each route, or, alternatively, so as to replicate the average distance travelled and average population weight for the system. In this latter formulation, the model produces expected levels of interaction on each route based purely on distance and population levels, and meeting all of the constraints imposed (such as the total expected interactions being equal to the total observed interactions). This second approach was selected for this study, since the aim in calibrating the model is to produce estimates of traffic potential, given the node characteristics, rather than try to reproduce actual interaction levels.

Since the data provided by Continental arbitrarily indicated one center as an origin and the other as a destination, use of an origin-constrained model was rejected, as no account would be taken of capacity constraints on the destination nodes. In addition, as was pointed out previously, the singly-constrained models are spatially biased whereas the doubly-constrained models take into account the relative spatial location of nodes [Fotheringham (1983)]. For these reasons, the doubly-constrained model was thought to be superior and the interaction levels produced were the closest to the "network potential" attribute that was needed in the production function analysis. The model is given as

$$T_{ij} = Z_i \, O_i \, B_j \, D_j \, W_{ij}^{\Theta 1} \, d_{ij}^{\Theta 2} ,$$

where

$$Z_i = \Big(\sum_{j=1}^{n} B_j \, D_j \, W_{ij}^{\Theta 1} \, d_{ij}^{\Theta 2} \Big)^{-1}$$

and

$$B_j = \Big(\sum_{i=1}^{m} Z_i \, O_i \, W_{ij}^{\Theta 1} \, d_{ij}^{\Theta 2} \Big)^{-1}.$$

T_{ij} is the proportion of interactions on route (ij). W_{ij} is the weight for the route, given by the combined effect of the populations and market shares at i and j. The d_{ij} term gives the intervening distance between i and j. O_i is the total passenger traffic at origin i, while D_j is the total passenger traffic at destination j. Z_i and B_j are balancing factors ensuring that the known total flow out of origin i and into destination j equal the estimated total inflows and outflows. The thetas are the parameters to be estimated.

The doubly-constrained gravity model was initially calibrated by using the first order iterative method of Hyman and Evans [see Batty and Machie (1978)]. The convergence of this process was good in the initial iterations, but after 30 or more iterations, the process moved very slowly towards the convergence limit. The parameter estimates produced were relatively inefficient because the convergence limit could not be reached.

An adjustment to the first order process was produced by Feighan (1989), whereby parameters were adjusted in successive iterations by examining the sign and extent of the proportional improvement in the previous adjustment's movement towards the target, rather than simply looking at the ratios of previous adjustment values to the target. This process converges quickly and met the convergence limit of 2% within the target value.

The parameter values produced using this method, (or indeed any other method), depend on the magnitude and range of the W_{ij} and d_{ij} values. It was decided to standardize both variables and scale them so that they lay on the interval [0,10]. This means that both the distance variable and the adjusted population or weight variable were seen as being equally important in their effects on the interaction levels. The original variables covered very different ranges, with the weight variable on average being of the order of 10^5, while the distance variable was of the order of 10^3. In addition to giving these variables equal weight in the relationship with interaction levels, it is argued that such a scaling makes comparisons with other models possible and valid. The results are summarized in Table 5.9.

The most striking result in the table is the negative parameter value generated for the weight variable. Different starting values were selected and the calibration re-run, but in every case the weight parameter approached the value shown in the table. The distance parameter was always negative, as expected.

The reasons suggested for the negative parameter values are two fold: (1) the parameter represents Continental's selection of a small set of the larger FAA hubs as nodes of concentration in their system, while weight levels at many of the other nodes would produce expected levels much higher than observed. Similarly at

Table 5.9
Convergence of doubly-constrained gravity model by adjusted first order process.

Convergence criteria:
(1) Distance: observed mean trip length ($\mathbb{C}$) in normalized and scaled units (d_{ij}) approximates estimated mean trip length for iteration m ($\hat{\mathbb{C}}_m$) where

$$\mathbb{C} = \Sigma\, \Sigma\, p_{ij}\, ln\, d_{ij} \quad \text{and} \quad \hat{\mathbb{C}}_m = \Sigma\, \Sigma\, \hat{p}^m_{ij}\, ln\, d_{ij}$$

where p_{ij} is the proportion of total interactions, and $\hat{p}^m_{ij}$ is the estimated proportion at iteration m.
(2) Weight: observed mean capacity ($\mathbb{B}$) in normalized and scaled units (W_{ij}) approximates estimated mean capacity at iteration m ($\hat{\mathbb{B}}_m$) where

$$\mathbb{B} = \Sigma\, \Sigma\, p_{ij}\, ln\, W_{ij} \quad \text{and} \quad \hat{\mathbb{B}}_m = \Sigma\, \Sigma\, \hat{p}^m_{ij}\, ln\, W_{ij}\ .$$

Results of calibration

Iteration of the model	Distance $-\theta_1$	$\mathbb{C} - \hat{\mathbb{C}}_m$	Weight θ_2	$\mathbb{B} - \hat{\mathbb{B}}_m$
1	2.000	-.09	-2.000	-.18
5	1.584	-.04	-4.170	-.09
10	1.270	-.02	-5.822	-.03
15	1.261	-.04	-6.638	-.02
20	1.075	-.02	-6.638	-.02

the other end of the urban hierarchy, a number of smaller centers receive considerably more traffic than would be expected going purely by population size and market share. (2) The adding-up or balancing factors produce higher proportions of expected traffic at the hub nodes, and the weight variable adjusts downward the proportion on the routes. Essentially then, the parameters are a function of the model structure and the breakdown in the urban hierarchy to interaction ratios mentioned in (1).

To further make the point, by examining the U.S. Census rankings of the SMSAs by population in 1980, it may be seen that the top five centers are New York, Los Angeles-Long Beach, Chicago, Philadelphia and Detroit. Continental served each of these nodes in 1987/88, but service levels were well below comparable levels at the 9th ranked Houston, the 17th ranked Newark and the 22nd ranked Denver. This pattern would be observed for every airline in the U.S., since the propensity to hub by airlines requires selection of one, two or three key nodes as the focus of their network system. This produces regional specialization or concentrations by airlines, which often bear little resemblance to the urban hierarchy _in isolation_. Fotheringham (1983) demonstrated the strong gravity relationship for the Continental U.S. for total air traffic, but this research shows how the single airline's traffic pattern does not follow the same trend with population levels even when market share is taken into account.

The doubly-constrained gravity model produced expected traffic proportions on each of Continental's routes. These proportions were converted to seats available per departure by multiplying the proportions by the total number of seats available per indexed number of departures. As with all of the variables in the production models, this variable was then scaled by the normalized number of departures on the route.

This chapter described in detail the data that was available and that was generated for the production analysis that is discussed in the next chapter. The extensive data set given by Continental Airlines was described using a series of cross-tables to highlight where differences in characteristics such as load factors, connection rates, share of departures and types of equipment occur in the system. The classifications used in this descriptive analysis will be used again in the next chapter so that a sense of perspective can be maintained in the discussion of efficiency.

The variables generated for the production function analysis were then described. One problem that can occur with generated data is that the variables may be highly correlated with one another, and this situation arose for four of the input factor variables. This problem was solved by using Principal Components Analysis to produce a single factor capturing most of the common variability in the four inputs. It was shown that this factor could be justifiably called an 'equipment' variable, since the four original variables, pilot and flight attendant hours, fuel and vehicle capital, depended directly on the choice of equipment.

The new network variable being included in the production model was generated from a doubly-constrained gravity model. This gravity model had a weight variable which captured the joint effects of the adjusted node populations for the route (the adjustment took account of the number of airports in the cities/towns) and the airline's share of traffic at each node. A distance variable was also included and traffic potentials were derived for each route.

This set of variables was used to estimate the production functions for the system of routes, and for four subsets of routes served by Continental Airlines in 1987/88. In the next chapter the results are presented along with an analysis of the efficiency characteristics derived from the frontier function procedure.

CHAPTER 6

PRODUCTION FUNCTION ANALYSIS OF CONTINENTAL AIRLINES

In this chapter, the results of the production functions estimated for Continental Airlines' domestic route system are reported. The Chapter is divided into two sections. In section 6.1 the results of the Cobb-Douglas production function estimates are presented and discussed. Five functions were estimated in all, and these functions are compared in section 6.1 to highlight differences in the organization and performance of different hub subsystems and non-hub routes. The section presents strong empirical evidence for including a network variable in the airline firm's production function.

In the second section the results of the frontier function estimates are presented and analyzed. The characteristics of efficient and inefficient routes in the system are set out in a series of tables. These tables are in the same format and use the classifications that were used in Chapter 5 to describe the route characteristics. This allows for comparison of a wide range of features.

6.1 Cobb-Douglas Production Function Estimates for Continental Airlines in 1987/88

Table 6.1 records the results of the GLS estimation procedures for the Cobb-Douglas production function, for the system of domestic routes and for the four subsets of routes, namely the Denver, Houston and Newark hubs, and the non-hub routes. With just four main variables, the AGENT, LOAD, NETWORK and EQUIPMENT variables, the goodness of fit was very high for the system of routes at R^2= 0.885, it being significantly higher for the three hub subsets at an average of 0.98. The goodness of fit for the non-hub routes was lowest at 0.94. The network variable was significant for the system equation and also for the Denver and Houston subsets at the 95% level. It was significant at 94% and 92% respectively for the Newark and nonhub equations. The load factor variable is significant in all equations except the nonhub equation. The equipment variable is significant in all equations.

The coefficients for the original variables (PILOT, FA, FUEL and CAPITAL) were computed using the factor loadings from the Principal Components Analysis (see Chapter 5) and are recorded in Table 6.2.

In order to make a comparison of the influence of each variable, and in order to discuss the returns to scale, the standardized coefficients were computed and recorded for all variables in the five groupings of routes. These are given in Table 6.2. The standardized coefficients were necessary because they removed the influence of different units of measurement in the explanatory variables. The load factor for instance was measured on a scale of 0 to 1, and the coefficients in this case particularly needed to be standardized.

Looking at the coefficients in the equations in more detail now, the network factor suggested that in Continental's domestic system generally, a 10% rise in passenger potential will lead to a 1% rise in output levels as measured by passenger seats utilized per indexed departure. The coefficient is higher for the hub routes, indicating that the passenger potential required to increase output levels on hub routes is much lower than that required to increase output levels on the non-hub routes. Confidence intervals were constructed around the estimated coefficients, and these intervals together with t-ratios are recorded in Table 6.3. It should be remembered that this network variable represents the aggregated influence of (i) the populations of the two airport communities comprising the route (adjusted for the number of airports in the municipality) (ii) the airline market share on the route and (iii) the length of the route. The significance of the NETWORK coefficient is an important finding in supporting the theoretical arguement for inclusion of a network variable in a production (and cost) model.

The load factor coefficient is negative for the system of routes. For the Denver, Houston and Newark routes, this capacity variable is positive and significant, indicating that greater density of traffic on hub routes increases output levels by between 10% and 15%. While the airline looks on high load factors as positive assets measuring passenger density on routes, the load factor may be seen as a quality variable from the passenger's

Table 6.1

GLS estimates of Cobb-Douglas production function coefficients for Denver, Houston, Newark and Nonhub and the full system of domestic routes.

Dependent variable = average passengers per departure, indexed by the number of departures per route.
(Standard errors in parentheses)

Independent variables	Denver	Houston	Newark	Nonhubs	System
AGENTS	.396 (.054)	.459 (.018)	.489 (.216)	.331 (.020)	.232 (.077)
LOAD	1.010 (.135)	.796 (.185)	1.205 (.277)	-.150 (.155)	-.331 (.108)
NETWORK	.071 (.019)	.056 (.023)	.094 (.056)	.028 (.025)	.079 (.021)
EQUIPMENT[1/]	-.237 (.028)	-.270 (.048)	-.249 (.086)	-.272 (.051)	-.307 (.036)
CONSTANT	2.207 (.084)	1.887 (.111)	2.045 (.168)	0.552 (.164)	1.085 (.825)
$\bar{R}^2$ [2/]	.986	.983	.973	.940	.885
SEE[3/]	.208	.347	.793	.796	.587
RHO	.087	-.333	-.134	.050	.022
DF[4/]	75	56	46	100	291

1/ EQUIPMENT variable is the first principal component for the input variables fuel, pilot, flight attendant and capital quantities per departure, weighted by the number of departures per route.
2/ Adjusted R-squared
3/ Standard error of the estimate
4/ Degrees of freedom

point of view, and higher load factors suggest a lower quality of service. For the nonhub routes, the load factor coefficient is negative, though not significant. Thus a higher load factor would make the passenger less inclined to travel on a nonhub route with this airline. Routes not directly connected to Continental's three main hubs have generally lower load factors than the hub routes, but these nonhub routes have higher connection rates. It is apparent that a different set of decisions and tradeoffs need to be made by the airline in the case of 'spoke' routes. This will be further discussed in the next section on efficiency.

Table 6.2

GLS estimates of Cobb-Douglas production function coefficients for Denver, Houston, Newark, Nonhub and the full system of domestic routes, in standardized units.

Dependent variable = average passengers per departure, indexed by the number of departures per route.

GLS estimates.

Independent variables	Denver	Houston	Newark	Nonhubs	System
AGENTS	.364	.416	.443	.392	.243
LOAD	.152	.103	.145	-.041	-.074
NETWORK	.102	.078	.113	.040	.105
FUEL	.108	.123	.090	.117	.138
PILOT	.114	.118	.083	.143	.155
FA	.131	.136	.100	.150	.174
CAPITAL	.121	.129	.082	.189	.187
RTS*	1.091	1.101	1.056	.989	.928

* Overall Returns to Scale

Table 6.3

95% confidence intervals for the network variable in the GLS estimates of Cobb-Douglas production functions.

Equation	Lower bound	Coefficient	Upper bound	T-ratio	Probability
Denver-GLS	.033	.071	.109	3.72	.000
Houston-GLS	.010	.056	.102	2.44	.000
Newark-GLS	-.018	.094	.206	1.67	.060
Nonhub-GLS	-.021	.028	.078	1.13	.080
System-GLS	.037	.079	.121	3.70	.000

The agent or ground staff input is positive and significant in all equations. The coefficients are higher for Houston and Newark routes. For the system as a whole, the coefficient indicates that a 10% increase in passenger levels per weighted departure require an equivalent rise in agent/ground staffing levels of 2.4%. Newark's routes require a 4.5% increase in staffing levels, presumably highlighting the airline's need to establish itself at this hub and better integrate passenger flows with the rest of its system. At the two more established hubs, the relative staffing levels are slightly lower.

The equipment variable has a negative sign in all equations because this variable is measured in principal component deviation units. The transformed coefficients on the original four variables are included in Tables 6.2 and 6.3. The coefficients show little variation within and between equations, and this variation can be explained by the PCA factor loadings.

For the non-hub routes, the capital coefficient is the most substantial, being 7% greater than the fuel coefficient. The lower fuel coefficient reflects the lower load factors on these routes. For the Denver, Houston and Newark routes, the flight attendant coefficients have a slightly higher weighting in the production process than the other equipment variables (and this corresponds to the factor loadings). The four variables collectively make up the major input component in the production of flight services, as expected. Variability among the coefficients relate directly to variation in load factor and distance-time ratios, since, in computing the variables, these two factors were the greatest influence in pro-rating the variables.

The overall returns to scale are made up of route-specific returns to scale and economies of scope. Looking at these overall returns to scale for the system of routes, the GLS estimates indicate that, in general, the domestic system is characterized by approximately decreasing returns to scale (RTS)= 0.928. This means that on average, an increase in inputs will be matched by a less than equivalent increase in output, taking into account possible economies of scope and route specific economies of scale. Looking at the subsets of routes however, Table 6.2 highlights how the three hub route systems are characterized by slightly increasing returns to scale, while the spoke routes indicate that returns are decreasing. For the hub routes, overall RTS vary from 1.08 for Denver to 1.12 for Newark. Thus, an increase in the scale of operation at any of the hubs, as measured by the levels of inputs used, will lead to a more than equivalent increase in output levels. The hub routes handle far greater volumes of traffic, have a greater range of equipment types available, and generally are of greater stage length than the spoke routes. The increasing returns to scale associated with each of the hub subsystems encourages the airline to expand the number of connections to hub airports and discourages from adding to the set of 'spoke' routes which are not directly connected with one of the hubs. The sources of scope economies can best be realized under these circumstances. Presumably, the airline's control of substantial proportions of gate and terminal space contributes to the increasing scale returns.

The nonhub routes have lower load factors, and require a greater proportion of 'equipment' inputs, than do the hub routes. Given that these routes are generally of shorter stage length, the decreasing returns highlight the relative inefficiency of jet equipment assigned to these routes. In addition, a greater increase in passenger potential is required on non-hub routes to bring about a standard increase in output levels.

In the next section, the production performance of each route will be measured by comparing the actual output levels achieved with 'best practice' output levels.

6.2 Measuring the Efficiency of Continental's Route System, and Exploring the Characteristics of Inefficient Routes

Frontier production functions were estimated for the system of routes using Corrected Ordinary Least Squares (COLS) and Maximum Likelihood estimateion with an assumed Gamma distribution for the error term (MLEγ). The two methods produced very similar results in terms of the measurement of efficiency of each route in the system. Rather than compare frontier output and observed output in average passenger seats per indexed departure units, it was decided that measuring efficiency in standardized efficiency units would be a more useful exercise, and that comparison of results would be easier across subsets of routes. In addition, the frontier function depends to a large extent on the position of the positive outliers. The most efficient route in the system was an infrequently used route, with a very high load factor and low connection rates. Comparison of all other routes to just this route would be misleading.

The frontier output level was compared with the observed output level, and the mean difference (i.e. the mean level of inefficiency) was computed along with the standard deviation of the difference. For each route, the level of inefficiency was then converted into standard normal deviates. In the tables and discussions that follow, efficiency is thus measured in standard deviation units around the mean level of inefficiency.

For the system of routes, Tables 6.4 and 6.5 detail the average inefficiency for routes, categorized in distance and equipment-type hub groupings and also in 'range of equipment-type' classes. In addition, inefficiency levels were grouped into 4 categories so that load factor and connection rate comparisons could be made. The groupings were (1) most inefficient routes, with average inefficiency greater than -0.25 standard deviations from the mean, (2) below average, inefficient routes, with average efficiency between -0.25 standard deviations and the mean inefficiency, (3) average inefficient routes, with average inefficiency between the mean and 0.5 standard deviations above the mean, (4) most efficient routes, with average inefficiency greater than 0.5 standard deviations above the mean. This scheme represents the influence of the Gamma distribution in the estimation procedure, and it also demonstrates the small amount of variation that was present in the system of routes.

For both COLS and MLEγ estimates, hub routes had lower inefficiency levels than non-hub routes. Among the hub routes, Newark routes were the most efficient with average inefficiency above the mean level. Houston-based routes were slightly more efficient on average than Denver routes, though over shorter distances, routes directly connected to Denver were more efficient in producing flight services, than either Newark or Houston. In general, the routes with an average stage length of at least 1,000 miles were the most efficient in the system.

There was a strong trend towards increasing levels of inefficiency, as routes had progressively more choices of equipment types available. Over all distance categories, routes using between one and three different aircraft types were more efficient than routes with greater aircraft pools. However, such routes accounted for less than 5% of total departures in 1987/88.

Looking at the different types of equipment, (i.e. long-, medium- and short-haul aircraft), there was little variation in efficiency levels for the three different groups. Average load factors and connection rates were computed for the four 'inefficiency classes', and compared over different distance and market categories. The most efficient routes had the highest load factors in all distance categories.

The opposite trend was observed for the connection rates; generally, routes which were the most inefficient in the system had the highest connection rates. The most efficient routes had an average connection rate of 49.8%, when the system average was 52.1%. Tables 6.6 and 6.7 summarize these trends for the system. For Continental's system of routes it must be noted that the longer distance routes had higher load factors and lower connection rates. The efficiency measure suggests that stage length and presumably distance-time ratios are more important influences on the firm's production levels. Alternatively, it could be said that when load factors are relatively stable, distance becomes a more important influence. In Chapter 5 it was shown that the non-hub routes had the highest connection rates and lowest load factors and that these routes have now been shown to be the least efficient in Continental's system. In section 6.1, these routes were seen to be characterized by decreasing returns to scale (overall). The utility of these routes to the airline lies in the fact that they very much boost traffic levels on the hub-system routes.

In summary then, for the system of routes, connection rates are negatively correlated with the level of efficiency, while distance, load factor and lower equipment levels are positively correlated. The three major hubs in the system are more efficient on average than the 'spoke' routes, although among the hubs, the smaller hubs are more efficient in producing flight services. In order to find out why it was that Newark, the least 'hub-like' node in Continental's system, should be the most efficient,

Table 6.4

Average inefficiency, measured in standard deviation units from COLS regressions, for crosstables of distance categories with groups of aircraft range,[1/] amounts of different equipments available,[2/] and Continental hub airport markets

Distance in miles	Long range	Medium range	Short range	Total
0 - 249	-.067	-.208	-.202	-.199
250 - 499	.341	-.090	-.080	-.070
500 - 749	-.089	-.035	-.011	-.028
750 - 999	-.116	-.046	-.023	-.039
1000 -1499	-.074	.007	.274	.087
1500 -2565	-.019	.005	-.021	-.005
Total	-.016	-.071	-.019	-.049

Distance in miles	Number of different equipment types available			Total
	1 - 3	4 - 6	7 - 10	
0 - 249	-.168	-.193	-.207	-.199
250 - 499	.119	-.106	-.068	-.070
500 - 749	.237	-.023	-.066	-.028
750 - 999	.229	-.007	-.099	-.039
1000 -1499	2.679	-.042	-.121	.087
1500 -2565	.046	-.012	-.062	-.005
Total	.497	-.063	-.111	-.049

Distance in miles	Denver	Houston	Newark	Interhub	Other	Total
0 - 249	-.143	-.158	-.142	0.000	-.244	-.199
250 - 499	-.077	-.112	-.117	0.000	-.019	-.070
500 - 749	-.036	-.086	.115	0.000	-.061	-.028
750 - 999	-.057	.065	-.170	-.092	-.061	-.039
1000 -1499	-.042	-.032	.618	-.079	-.129	.087
1500 -2565	-.096	-.017	.052	-.056	.090	-.005
Total	-.059	-.046	.070	-.074	-.117	-.048

Notes

1/ Aircraft ranges are defined long haul, medium haul and short haul according to _Jane's All the World's Aircraft._

2/ Number of different equipment types gives the number of different aircraft serving a route in 1987/88.

Table 6.5

Average inefficiency measured in standard deviation units for MLEγ regressions, for crosstables of distance categories with groups of aircraft range, amounts of different equipments available, and Continental hub airport markets

Distance in miles	Long range	Medium range	Short range	Total
0 - 249	-.033	-.078	-.076	-.075
250 - 499	.194	-.069	-.077	-.062
500 - 749	-.133	-.071	-.060	-.069
750 - 999	-.164	-.084	-.073	-.083
1000 -1499	-.153	-.029	.230	.044
1500 -2565	-.133	-.101	-.106	-.109
Total	-.089	-.069	-.025	-.054

Distance in miles	Number of different equipment types available			Total
	1 - 3	4 - 6	7 - 10	
0 - 249	.036	-.011	-.132	-.075
250 - 499	.342	-.066	-.092	-.062
500 - 749	.193	-.060	-.110	-.069
750 - 999	.253	-.055	-.143	-.083
1000 -1499	2.715	-.100	-.161	.045
1500 -2565	-.074	-.116	-.143	-.109
Total	0.520	-.062	-.126	-.054

Distance in miles	Denver	Houston	Newark	Interhub	Other	Total
0 - 249	-.114	-.109	-.103	0.000	-.047	-.075
250 - 499	-.120	-.115	-.098	0.000	.015	-.062
500 - 749	-.079	-.118	.039	0.000	-.079	-.069
750 - 999	-.119	.000	-.165	-.172	-.046	-.083
1000 -1499	-.114	-.114	-.634	-.166	-.128	.045
1500 -2565	-.158	-.115	-.055	-.151	-.105	-.109
Total	-.111	-.085	.066	-.162	-.045	-.055

individual routes were examined for various operating characteristics. The COLS and MLEγ estimates both weighted the Cleveland-Newark route heavily, so that this route was substantially more efficient than any other route in the system. The standardized efficiency level for this route was 16.53 units above the mean inefficiency level. Clearly then, Newark's overall average statistics reflect the heavy influence of this extreme value. Removing this route and recomputing average inefficiency for Newark's remaining routes produced a relatively similar value to Denver and Houston's averages. This outlier obviously influenced the other totals in the tables too. The Cleveland-Newark route was characterized in 1987/88 by moderate stage length, with a very high load factor and low connection rate.

Table 6.6
Percentage of departures in each 'efficiency' group by amounts of different equipments available.

Efficiency in σ units	Number of different equipment types available			Total
	1 - 3	4 - 6	7 - 10	
<-.25	0.271	6.070	12.608	18.950
-.25- μ	2.397	17.571	36.665	56.634
μ - .50	0.271	2.739	8.038	11.048
> .50	1.628	6.209	5.531	13.368
Total	4.568	32.590	62.842	100.000

Table 6.7
Average connection rates for efficiency groups over subsets of Continental's route system

Efficiency in σ units	Denver	Houston	Newark	Interhub	Other	Total
<-.25	0.717	0.589	0.225	0.000	0.723	0.608
-.25 - μ	0.593	0.560	0.326	0.464	0.516	0.516
μ - .50	0.551	0.549	0.268	0.000	0.446	0.440
> .50	0.570	0.513	0.253	0.000	0.510	0.498
Total	0.592	0.552	0.290	0.464	0.580	0.521

The main conclusions that can be drawn from this analysis are (1) the hubbing of routes is more efficient for the firm than point to point operations, or 'nonhub routing'. (2) For all routes, a better matching of equipment types with stage length reduces the pool of aircraft available to the route, and this is more efficient. (3) For the nonhub routes, presumably the high connection rates have an important role in making the shorter routes more efficient: For the system generally, the longer segments are the most efficient, and when high load factors can be achieved also, a substantial improvement in efficiency can be observed. For Continental Airlines in 1987/88, the company was attempting to integrate the networks of the three airlines which were merged in 1983/84 under the reorganized Continental banner. The company had generally old equipment and while it is acknowledged that the hub-and-spoke routing strategy lead to a more efficient network organization, Continental needs to reorganize its capital stock and plan new equipment purchases which better match route characteristics with passenger demands. The method used in this production analysis could be extended to test for the performance of new types of equipment or, different equipment pools, on routes in the system.

In the final chapter of this book, the main findings reported in this chapter are summarized along with the main conclusions from earlier chapters. The final chapter makes a series of recommendations concerning future studies of the air transport industry at the micro-level and also at the industry level. This chapter has shown how inclusion of a network variable and route-interdependence was supported for a large U.S. airline.

CHAPTER 7
CONCLUSIONS AND RECOMMENDATIONS

In this chapter, a brief summary is provided of the main findings from the U.S. air system analysis of Chapter 3, and the individual airline analysis of Chapters 5 and 6. Throughout these reviews, further avenues of research will be pointed out, and in the case of the airline-level study, recommendations are made concerning further empirical work.

7.1 Network Changes and the Role of Network Organization

In Chapter 2, it was shown how recent work in the economics literature did not take adequate account of the important role of the airline's network organization, when assessments of cost improvements and productivity gains were being made. In many instances contradictory results were presented and it was suggested that the exclusion of a variable capturing the key network component of the airline's operation may explain these discrepencies. The inclusion of this component would very much enhance the quality of the models and make them more realistic. While a small number of studies did include crude indicators of an airline's network characteristics (such as the number of points served, the Herfindahl index or the proportion of traffic at the airline's major hubs), it was argued that these measures did not adequately capture the manner in which the network was organized.

Later in Chapter 2, the many details of concern to schedule planners were reviewed, and it was clearly shown that network organization was of critical importance in the firm's operation because it defined not only the production plan, but also the product. Motivated by a dissatisfaction with existing models of airline production and costs, and inspired by the view of the airline put forward by operations researchers, it was decided to undertake new research and focus on two main themes. These were: (1) to investigate the nature and extent of airline network changes over a lengthy period so that comparisons could be made in relation to the effects of deregulation. The impact of these network changes on the airport system had to be assessed also. (2) to develop a production model of the individual airline firm that included the essential features of the network and the organization of that network. The model was developed in such a manner as to allow for assessment of production performance on a route by route basis.

In Chapter 3, the actual changes in traffic patterns in the U.S. generally, the changes at different categories of airports, and the changes in airline traffic patterns were examined in detail. It was shown how traffic levels had dramatically increased since 1978, and how the national airway system supporting that traffic had contracted. Thus airports in the U.S. receiving scheduled traffic service had undergone either considerable expansion or contraction since the passage of the ADA.

Major air carriers in the U.S. have greatly expanded their network size, either by internal growth or mergers/takeovers of other carriers. The networks of the large carriers are very much focused on a small series of key nodes or airline hubs, and the tendency to streamline traffic through these nodes has become much more pronounced in the 1980s. The Gini index was shown to be a superior measure of network concentration, compared with the measures used previously in the economics literature on the airline industry. This measure takes account of the traffic allocation at every point in the airline's system and accordingly compares traffic levels at hub airports with smaller markets in indexing the concentration level. For economically intuitive and theoretical reasons, this index was proposed as a superior measure of network concentration. It is recommended that in future analysis of the airline industry, relative concentration measures, such as the Gini index and Theil's measure, be used instead of the measures proposed by Kanafani and Hansen (1985), McShan (1986) and others referred to in Chapter 2.

The Gini index showed clearly how airport and airline concentration levels were very high prior to deregulation, so that the hubbing phenomenon recently observed, had been in place for a long period before passage of the ADA. Deregulation allowed this pattern of concentration to become more pronounced, but it did not create it.

For macrolevel analysis of U.S. commercial air traffic at FAA hubs, the decomposition scheme proposed for the Gini and Theil measures in the literature on income distribution could be used to analyze specific airline's, or groups of airline's influence on the national system, as they have re-organized and concentrated their networks. This kind of analysis would be of particular value to regional airport planners, or federal planners, interested in forecasting or planning for further growth in the 1990s.

The airport and airline industry analysis of Chapter 3 showed clearly that the production plan and the products offered by airlines after 1978 had substantially changed. The focus was then refined to a micro-scale so that the precise form of this network technology could be included in an economic model of the airline.

7.2 Airline Level Analysis and Route Efficiency

In Chapter 4, the focus of attention was reduced to the firm level and the airline's decision-making process was set out in a production function model. It was assumed that the airline made three key decisions in selecting its network and its output levels across the system of routes. First of all, the firm decides on the airports to serve and this is usually historically determined so it was assumed that these airports are given and move on to the next step. The second decision involves deciding how these airports should be linked. This problem was solved using the famous 'Transportation problem of Linear Programming' extended to include a greater degree of accessibility in the optimal network. The final decision is then to decide on the service level to produce on each route and this can be summarized in a production function adjusted to include the traffic potential for each route. These traffic potentials are generated from the solution to the transportation problem. The empirical model was derived in Chapter 4 and shown to include the interdependence of routes by using a spatial autoregressive estimating model, where connection rates among routes captured the joint-production effects. This was required in order to capture the economies of scope associated with production of transport services over a network.

In Chapter 5 the data used in the production analysis was

described in detail. This consisted of an extensive route-by-route listing of Continental Airlines' domestic operation for the year 1987/88, data from published sources and data generated from these two compilations. The input and output variables were then used to estimate production functions and frontier production functions, the results of which were given in Chapter 6. This chapter clearly highlighted the role of the network and the connection rates (representing network integration at the system level) in explaining variations in output levels across one firm's system of routes.

The development of an appropriate network variable was also detailed in Chapter 5. A number of different measures of network potential were investigated. For the system of routes, gravity model predictions gave an expected level of interaction on all routes, based on Continental's market share and population levels at the city pairs, as well as the intervening distance. In addition, seating capacities rather than actual seating levels were used to calibrate the model.

Further investigations could be undertaken in this area, in terms of widening the scope of the network potential concept. For example, populations could be stratified by urban functions, so that a better understanding of passenger markets could be incorporated into the models. At the macrolevel, Fotheringham's model needs to be calibrated for the Continental U.S. using up-to-date origin-destination data for the industry. Individual carriers could then calibrate their traffic potential and market shares using Fotheringham's specific parameters on accessibility levels at all airports within and indeed outside of their route systems. This essentially means including potential demand for service in a production model for the existing network and for hypothetical extended networks.

It was explained why route-level analysis was more appropriate than airport-level analysis at the micro-scale. This was principly because the essential feature of the airline firm was its production of flight services on a set of routes rather than from a number of airports. The route analysis captured the technology embodied in network organization, while airport-level analysis ignored this important effect. It was pointed out that there were potential statistical and intuitive problems with analysis at the airport level.

In Chapter 6, the production function analysis supported the inclusion of the network potential variable in the model; for the system of routes and for the hub routes, the network variable was positive and statistically significant. It was suggested that a more thorough model would include a full connection matrix for each pair of routes in the system. This type of analysis would be of greater interest to airlines since they would have ease of access to microlevel data, and because they could make shorter term (i.e. less than annual) assessments of each route's performance vis-à-vis the system of routes.

The investigation of efficiency in the network system revealed the difficult tradeoff needed in matching equipment levels and passenger demand, and also route traffic volumes and frequency of service offered. Those routes of greater stage length and utilizing a smaller range of equipment types were relatively more efficient for the airline in 1987/88. But this subset of routes accounted for a small proportion of the total system's volume of traffic in that year; on the heavier routes, it can be presumed that offering a consistent level of daily service reduces the average load factor for the year, since seasonal variations and daily variations in traffic levels will require a greater seat capacity allocation. The heavy route segments where Continental has a high market share are the back bone of the airline, and while feeder routes may be cut or added (or indeed service levels may be increased or decreased) with relative ease, on the heavier route a consistent level of service is necessary to channel traffic to and from the spoke routes. For these hub routes, a better matching of equipment with passenger demand and route length is the most important aspect that can lead to improved efficiency levels.

This kind of analysis needs to be undertaken with a full system of connections between each pair of routes, so that estimates are more accurate. In addition, actual data recorded for the equipment variable would very much improve the reliability of the estimates. Microeconomic analysis of this kind is useful and helpful in the schedule evaluation process, but it depends in the first place on good route level data. Efficiency studies could easily be adapted to shorter time frames or to subregional markets, once the connection to other hub systems are catered for. Using actual data, the more sophisticated flexible functions would allow for detailed investigation of input factor substitutability, and this information could be used to improve equipment and route matching in the system. The model could be used to investigate the performance of new aircraft types or labor combinations, which would help decision-making in the areas of crew scheduling and equipment planning.

The most important recommendation that this research makes is that the role of the network be included in further analysis. For a transportation industry, the spatial dimension to the product, and therefore to any research in that industry, is of crucial importance. Recognizing this dimension enriches understanding of the production process, and improves the quality of the models characterizing the industry.

REFERENCES

AGIFORS (Airline Group of the International Federation of Operations Research Societies) Symposium Proceedings, Various years.

AIGNER, D., C.A.K. LOVELL & P. SCHMIDT "Formulation and estimation of stochastic frontier production function models, Journal of Econometrics, 6, (1977), 21-37.

ANSELIN, L. Spatial Econometrics: Methods and Models, Kluwer, Dordrecht, 1988.

ATKINSON, A.B. "On the Measurement of Inequality", Journal of Economic Theory, 2, (1970), 244-263.

ATKINSON, A.B. The Economics of Inequality, Oxford: Clarendon Press, 1975.

BAILEY, E.E., D.R. GRAHAM & D.P. KAPLAN Deregulating the Airlines, Cambridge: MIT Press, 1985.

BAILEY, E.E., & A.F. FRIEDLAENDER "Market Structure and Multiproduct Industries", Journal of Economic Letters, 20 (1982) pp. 1024-48.

BATTY, M. & S. MACHIE "The calibration of gravity, entropy and related models of spatial interaction", Environment and Planning A, 4, (1972), 205-233.

BAUMOL, W.J., J.C. PANZAR & R. WILLIG, Contestable markets and the Theory of Industrial Structure, New York: Harcourt, Brace Jovanovich, 1982.

BAUMOL, W.J. "Contestable Markets: an uprising in the theory of industry structure", American Economic Review, 72, Sept, (1982), 1-15.

BAUMOL, W.J. & R.D. WILLIG, "Contestability: Developments since the Book" in Strategic Behaviour and Industrial Competition, edited by D.J. Morris, P.J.N. Sinclair, M.D.E. Slater & J.S. Vickers, Oxford: Clarendon Press (1986).

BERNDT, E.R. & L.R. CHRISTENSEN "The translog function and the substitution of equipment, structures and labor in U.S. manufacturing, 1929-68", Journal of Econometrics, 1, (1973), 81-113.

BORENSTEIN, S. "Hubs and high fares: dominance and market power in the U.S. airline industry", Rand Journal of Economics, 20(3), (1989), 344-365.

BUNGE, W. "Theoretical Geography" Lund studies in Geography, Series C, No.1. 2nd Edn, (1966) Gleerup.

CAVES, R.E. Air transport and its regulators, Cambridge Mass.: Harvard Univ. Press, 1962.

CAVES, D.W., L.R. CHRISTENSEN & M.W. TRETHEWAY "Productivity performance of U.S. trunk and local service airlines in the era of deregulation", Economic Inquiry, 21, July, (1983), 312-324.

CAVES, D.W., L.R CHRISTENSEN & M.W. TRETHEWAY "Economics of density versus economies of scale:why trunks and local service airline costs differ", Rand Journal of Economics, 15(4), (1984), 471-489.

CIVIL AERONAUTICS AUTHORITY Report to the House Committee on Interstate and Foreign Commerce, Airport Survey, 76th Cong., 1st. session, March 24, 1939, 1. (H.Doc.245).

CIVIL AERONAUTICS BOARD "The consequences of airline mergers since deregulation", CAB and DOT Report, Washington D.C., 1982.

CIVIL AERONAUTICS BOARD Airline Operating Costs and Performance Report, Washington, D.C., 1984.

CIVIL AERONAUTICS BOARD Implementation of the Provisions of the Airline Deregulation Act of 1978, Report to Congress: Washington D.C., 1984

CIVIL AERONAUTICS BOARD & U.S. DEPT. OF TRANSPORTATION, Airport Activity Statistics of the Certificated Route Air Carriers, Washington D.C, 1965 to 1987.

DEVANY, A.S. "The effects of price and entry regulation on airline output, capacity and efficiency", Bell Journal of Economics. 6, (1975), 327-45.

DOUGLAS, G.W. & J.C. MILLER "Quality competition, industry equilibrium and efficiency in the price-constrained airline market", American Economic Review, 64 (Sept), (1974), 657-669.

EADS, G., M. NERLOVE & W. RADUCHEL "A long-run cost function for the local service airline industry: an experiment in non-linear estimation", Review of Economics and Statistics, Vol. 51 (3), (1969), 258-70.

ERLANDER, S. "Accessibility, entropy and the distribution and assignment of traffic", Transport Research, 11, (1977), 149-153.

ETSCHMAIER, M.M. & D.F.X. MATHAISEL "Aircraft scheduling: The state of the art" Proceedings of the XXIV AGIFORS Symposium, Strassbourg, France, 1984.

ETSCHMAIER, M.M. & D.F.X. MATHAISEL (1985) "Aircraft scheduling: An Overview", Transportation Science, 19(2), (1985), 127-138.

FEDERAL AVIATION AUTHORITY Statistical Handbook of Aviation, Washington D.C., 1979 to 1986.

FEIGHAN, K.J. "Estimating gravity model parameters", unpublished computer program, Dept of Civil Engineering, Purdue University, 1989.

FØRSUND, F.R, C.A.K. LOVELL & P. SCHMIDT, "A survey of frontier production functions and their relationship to efficiency measurement", Journal of Econometrics, 13, (1979), 5-25.

FOTHERINGHAM, A.S. "A new set of spatial interaction models: the theory of competing destinations", Environment and Planning A, 15, (1983), 15-36.

GAILE, G.L. "Measures of Spatial Equality" in Gaile & Willmott (eds.) Spatial Statistics and Models, 1984, 223-233.

GASTWIRTH, J. "On the measurement of U.S. Income distribution", Review of Economics and Statistics, 54(2), (1972), 256-259.

GENERAL ACCOUNTING OFFICE Airline Competition: Fare and Service Changes at St. Louis since the TWA-Ozark Merger, Briefing report to Hon. J.C. Danforth, Committee on Commerce, Science and Transportation, U.S. Senate (RCED-88-217 BR) Washington, D.C., 1988.

GETIS, A. & J. FRANKLIN "Second order neighborhood analysis of mapped point patterns", Ecology, 63(3), (1987), 473-477.

GRAHAM, D.R., D.P. KAPLAN & D.S. SIBLEY "Efficiency and competition in the airline industry", Bell Journal of Economics, 14(1), (1983), 118-138.

GOLDBERGER, A.S. "Interpretation and estimation of Cobb-Douglas production functions", Econometrica, 36(2), (1968), 464-472.

HANSEN, S. "Entropy and utility in traffic modelling", in Transport and Traffic Theory, Buckley (ed.), A.H. & A.W. Reed, Sydney, 1974.

HERFINDAHL, O.C. "Concentration in the Steel Industry" unpublished Ph.D thesis, Columbia University, New York, 1950.

IPPOLITO, R.A. "Estimating airline demand with quality of service variables", Journal of Transport, Economics and Policy 15, (1981), 7-15.

JENG, C.-Y. "Routing strategies for an idealized airline network", unpublished Ph.D Dissertation, University of California, Berkeley, 1987, (UCD-ITS-DS-87-3).

JENG, C.-Y. "An Idealized Model for Understanding Impacts of KeyNetwork Parameters on Airline Routing", Paper presented at 67th Annual Meeting, Transportation Research Board, Washington D.C., 1988.

JOHNSON, R.J., Multivariate Statistical Analysis in Geography, London: Longman, 1978.

KAMP, J. Air Charter Regulation: A Legal, Economic and Consumer Study, Praeger: New York, 1976.

KANAFANI, A. AND A. GHOBRIAL "Airline hubbing: Some implications for airport economics", Institute of Transportation Studies, University of California, Berkeley, 1984, (UCB-ITS-WP-84-5).

KANAFANI, A. AND M. HANSEN "Hubbing and airline costs", Institute of Transportation Studies, University of California,

1985, Berkeley, (UCB-ITS-RR-85-12).

KEELER, T.E. "Airline regulation and market performance", Bell Journal of Economics, 3, (1972), 399-424.

KORAN, D.W. "The welfare effects of airline fare deregulation in the United States", Journal Transport, Economics and Policy 17, (1983), 177-188.

KUZNETS, S. "Economic growth and income inequality", American Economic Review, 45, (1955), 1-28.

MACDONALD, J.B. "Some generalised functions for the size distribution of income", Econometrica 52(3), (1984), 647-63.

MADDALA, G.S. Econometrics, Singapore: McGraw-Hill International, 1977.

MAGNANTI, T.L. & R.L. WONG "Network design and transportation planning: Models and algorithms", Transportation Science, 18(1), (1984), 1-55.

McSHAN, W.S. "An economic analysis of hub-and-spoke routing strategy in the airline industry", unpublished Ph.D Dissertation, Northwestern University, 1986.

MEYER, J.R., C.V. OSTER, I.P MORGAN, B.A. BERGMAN & D.L. STRASSMANN Airline Deregulation - The Early Experience, Auburn House, Mass., 1981.

MEYER, J.R. "Book review: Bailey, Graham & Kaplan's 'Deregulating the Airlines'", Rand Journal of Economics, 17(3), (1986), 461-466.

MOORE, T.G. "U.S. airline deregulation: Its effects on passengers, capital and labour", Journal of Law and Economics, XXIX, (1986), 215-217.

MORRISON, S. & C. WINSTON The Economic Effects of Airline Deregulation, Brookings Institution, Washington D.C., 1986.

MORRISON, S. & C. WINSTON "The Dynamics of airline pricing and competition", American Economic Association Papers and Proceedings, 80(2), (1990), 389-393.

PAGLIN, M. "The measurement and trend of inequality: A basic revision", American Economic Review, 65, (1975), 598-609.

PANZAR, J. "Regulation, deregulation and economic efficiency: The case of the CAB", American Economic Review, 70, (1980), 311-315.

QUINONES, W. "Airports battle airlines on right of access to community airfields", Airport Services Management, January, (1986), 34-36.

REISS, P.C. & P.T. SPILLER "Competition and entry in small airline markets" Paper presented at the Federal Trade Commission Conference, Washington D.C., 1988.

REYNOLDS, A.J." Measuring the impact of network changes by the airlines in the deregulated environment", unpublished Ph.D dissertation, University of Illinois, Urbana, 1989.

RICHMOND, J. "Estimating the efficiency of production", International Economic Review, 15, (1978), 515-521.

SCHMIDT, P. "On the statistical estimation of parametric frontier production function model", Journal of Econometrics, 19, (1976), 233-238.

SEN, A."Poverty: an ordinal approach to measurement", Econometrica, 44(2), (1976), 219-231.

SHALIT, H. "Calculating the Gini index of inequality for individual data", Oxford Bulletin of Economics and Statistics 47(2), (1985), 185-89.

SPRY, E.C. "Route economics - accuracy versus meaningfulness", Economics of Air Transportation, Proceedings of the 7th World Airport Conference, Institution of Civil Engineers, London, 1984.

TAAFFE, E.J. "Air transportation and United States urban distribution", Geographical Review 46, (1956), 219-238.

TAAFFE, E.J. "Trends in airline passenger traffic: A geographic case study", Annals of the Association of American Geographers, 49, (1959), 393-408.

TAAFFE, E.J. "The urban hierarchy: An air passenger definition", Economic Geography 38(1), (1962), 1-14.

TAYLOR, J.W.R.(ed.) Janes All The World's Aircraft, London: Janes, 1982.

TAYLOR, P.J. Quantitative Methods in Geography, Boston, Mass.: Houghton Mifflin, 1977.

TRAPANI, J.M. & C.V. OLSON, "An analysis of the impact of open entry on price and the quality of service in the airline industry", Review of Economics and Statistics, 63(1), (1981), 67-76.

VARIAN, H.R. Microeconomic Theory, New York: W.W. Norton, 1984.

VERLEGER, K. "Models of the demand for air transportation", Bell Journal of Economics, 3, (1972), 437-457.

WEINER, E. "Vanishing act", Flying, 114(4), (1987), 70-74.

WHITE, L.J. "Quality variation when prices are regulated", Bell Journal of Economics, 3, (1972), 425-436.

WILLIG, R.D., "Multiproduct Technology and Market Structure", American Economic Review, 69(2), (1979), 346-351.

WILSON, A.G. Entropy in Urban and Regional Modelling, London:

Pion, 1970.

WONG, R.T. "Introduction and recent advances in network design: models and algorithms", in Transportation planning models, M.Florian (ed.) Elsevier Science publishers, B.V. North Holland, 1984.

ZELLNER, A., J. KMENTA & J. DREZE "Specification and estimation of Cobb-Douglas production functions", Econometrica, 34(4), (1966), 784-795.

AUTHOR INDEX

MIX
Papier aus verantwortungsvollen Quellen
Paper from responsible sources
FSC® C105338

If you have any concerns about our products,
you can contact us on
ProductSafety@springernature.com

In case Publisher is established outside the EU,
the EU authorized representative is:
Springer Nature Customer Service Center GmbH
Europaplatz 3, 69115 Heidelberg, Germany

Printed by Libri Plureos GmbH
in Hamburg, Germany